THE CHESS MYSTERIES
OF THE ARABIAN KNIGHTS

Raymond Smullyan, a well-known mathematician and logician, is Oscar Ewing Professor of Philosophy at Indiana University and Professor Emeritus of the City University of New York–Lehman College and Graduate Center. He is the author of many books, including *Forever Undecided, To Mock a Mockingbird, The Lady or the Tiger?,* and *The Chess Mysteries of Sherlock Holmes* (all available in Oxford Paperbacks).

The
CHESS
MYSTERIES
of the
ARABIAN
KNIGHTS

Raymond Smullyan

OXFORD UNIVERSITY PRESS

1992

Oxford University Press, Walton Street, Oxford OX2 6DP

Oxford New York Toronto
Delhi Bombay Calcutta Madras Karachi
Petaling Jaya Singapore Hong Kong Tokyo
Nairobi Dar es Salaam Cape Town
Melbourne Auckland

and associated companies in
Berlin Ibadan

Oxford is a trade mark of Oxford University Press

First published 1981 as a Borzoi Book by Alfred A. Knopf, Inc., New York
First issued as an Oxford University Press paperback 1992

British Library Cataloguing in Publication Data
Data available
ISBN 0-19-286124-7

Printed in Great Britain by
Biddles Ltd.
Guildford and King's Lynn

CONTENTS

CONTENTS

CONTENTS

HOW TO SOLVE
THESE PROBLEMS

*When one has eliminated the impossible,
whatever remains, however improbable,
must be the truth.*
—SHERLOCK HOLMES

It is not necessary to be an experienced chess player to enjoy the problems in this book; all that is required is a knowledge of how the pieces move. Unlike conventional chess problems (White to play and mate in so many moves), these problems (excepting the few in the addendum) are studies in chess-logic (the technical term is "retrograde analysis"). Like those in the companion volume, *The Chess Mysteries of Sherlock Holmes,* these problems are concerned with the *past* history of a given game, not its future. For example, a position may be given and you may be told that a piece stands on a certain square, *but not be told what the piece is.* The task then is to determine the color and denomination of the unknown piece. Or, again, a position may be given in which there are two White queens on the board, and the problem is to find out, by "reasoning backwards," which of the queens is the original and which the promoted one! The variety of interesting questions that can be posed in retrograde analysis is quite remarkable, and many people who have little interest in conventional chess problems are intrigued by problems of this sort.

Here are some examples.

To begin with, in all the problems in this book, squares will be designated by letter and number. For example, in the position on the next page, the White king is on d1, the Black king is on f2, the White rook is on h1, the White bishop is on c1, and the four White pawns are on b2, e2, g2, and h2.

Now consider the position. We are given that it is White's move. The problem is: Has there been any promotion in this game?

At first glance, this may seem a hopeless task. What possible clue is there?

Here is the solution: Since we are given that it is White's move, that means Black has just moved. Obviously it was with the king, since this is the only Black piece now on the board.

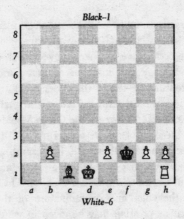

Black–1

From what square did Black come? Not from g3, since the pawn on h2 couldn't have moved to check him and therefore the king would have moved into check, which is not possible. Likewise, not from f3 because of the pawn on g2 (not to mention the pawn on e2). Could the king have come from e3? No, because he would have been moving out of check from the bishop on c1, and this bishop would have had no possible way to check him (because of the pawn on b2). Now, the Black king certainly couldn't have come from e1, because it is not legal for two kings to be in such proximity (the king that

x

moved last would have been moving into check). Therefore the Black king must have just come from f1 or g1, moving out of check from the rook on h1. But this raises a problem: How could White possibly have administered this check?

If the position now seems impossible, I'll give you a little hint: It is indeed impossible that the Black king just came from g1, but it is not impossible that he came from f1 (in fact, as you will soon see, he must have!). If you are still baffled, I'll give you another hint: couldn't the Black king—in moving from f1 to f2—have just captured a White piece on f2? So, going back a move, put the Black king on f1. Now, can't you put some White piece on f2 that makes the position possible? The answer is yes, a White bishop! This bishop just came from g1, discovering check from the rook, and then the Black king captured the bishop on f2. This is the only way the position could have arisen!

And so we have proved that right before the last move, there was a White bishop on f2. Now, f2 is a Black square, and since there is also a White bishop on the Black square c1, one of these two bishops must have been promoted from a pawn at some earlier stage of the game. So the answer to the question of the problem is *yes*.

Next consider a variant of the above problem: Remove the White pawn from e2. Now, it is given that Black moved last, but also that he *didn't* move from square f1! What was the move?

The answer is that the Black king just moved from e3, capturing a White rook on f2. Just before this, the rook had moved from d2, discovering check from the bishop.

Not all retrograde problems are solved by a consideration of the last move. Take the position shown on the next page.

It is given that one of the Black bishops captured a White piece earlier in the game. Which bishop was it—the one now on e5 or the one now on g4?

This time the most obvious fact is that the White pawn on c3 came from d2, capturing a Black piece on c3. This captured piece was a rook (which is the only missing Black piece). Now, because of the Black pawn formation, this Black rook couldn't have gotten out onto the board until *after* the Black pawn from g7 captured a White piece on h6. This piece couldn't have been the missing White bishop from c1, because before the capture on h6, the capture on c3 hadn't yet occurred. So the pawn on c3 was still on d2, and hence the missing White bishop was still locked in on its home square, c1. Therefore it is the missing White *knight* that was captured on h6. In other words, the sequence was this: First the Black pawn from g7

Black–15

White–14

captured a White knight on h6; then the Black rook got out via the square g7 and got captured on c3, and *then* the White bishop got out and was captured by a Black bishop—the one, of course, now on e5 (since the one on g4 travels only on White squares). So the answer is that it is the bishop on e5 that captured a White piece earlier in the game.

These problems are a bit like detective stories, and must be solved by finding the right "clues." I think my favorite type of retrograde problem is that in which a given position at first seems impossible, but then turns out to be possible after all.

The example below is a little reminiscent of something out of Edgar Allan Poe.

In this position, we are not given which side of the board is White and which side is Black, but we are given that no piece was captured on the last move. The problem is to determine which side of the board is White—the north side or the south side? (It couldn't be the east or west side, because the rules of chess demand that the lower righthand corner be a White square.)

Here is the difficulty: White's last move was obviously with

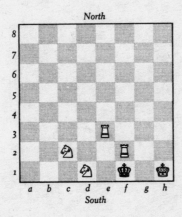

the rook now on f2; from what square did the rook come? If it came from any square on the f-file (from f3, say) it would have been checking the king before it moved, unless, of course, it had just captured a Black piece on f2. However, we are given that no piece was captured on the last move, and therefore the rook couldn't have come from any square on the f-file. This means that the rook came from one of the four squares h2, g2, e2, or d2. Now comes the puzzling part: Regardless of which of the four squares the rook came from, what could have been Black's move immediately before that? The Black king

couldn't have come from e1, because he would have been
moving out of check from both the rook on e3 and the knight
on c2. (Neither could have administered the check, because
the king would have already been in check by the other!) Also,
the king could not have moved from e2, since he would have
been simultaneously in check from both rooks. The king certainly didn't come from g1 or g2 (because of the White king on
h1). This leaves only the square f2, but how could the king
have come from f2 out of check from both the knight on d1
and the rook then on h2, g2, e2, or f2? So, regardless of which

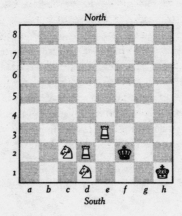

square the White rook came from, the position seems really
impossible.

But I'll give you a hint: The rook just came from d2. Does
that help? If not, I'll even tell you that Black's last move was
from f2—thus, the position before the last two moves was that
given above.

At this stage of the game, Black is simultaneously in check
from the rook on d2 and the knight on d1. Is this "double"
check really impossible?

No, it is not: the one possibility is that the North side is

White, and that White put Black in check by a pawn on e2 capturing a Black piece on d1 and promoting to a knight! Wildly improbable as this is, it is the *only* way the position could have arisen, and we recall Sherlock Holmes' words: "When one has eliminated the impossible, then whatever remains, *however improbable*, must be the truth." Therefore the North side must be White.

※ ※ ※ ※

Now you have some idea of what retrograde analysis is like, and as you can see, the ability to reason logically is the main requirement for understanding these problems. Of course, it is assumed that the reader is conversant with all the rules of chess; the rules governing castling, pawn promotion, and the *en passant* capture of pawns play a particularly prominent role in retrograde analysis. The beginning reader is hardly expected to solve the first few problems on his own; he will most likely have to resort to the solutions at the end of the book. But he will be surprised at how soon he will get the knack of things, and by the time he has finished this book, he will look back on the preceding problems as child's play.

CAST OF CHARACTERS

Haroun Al Rashid	*The White king*
Amelia	*His queen*
Kazir	*The Black king*
Medea	*His queen*
Olaf	*A knight*
Barab	*A Black pawn*
Gary	*A White pawn*
Archie the bishop	*Grand vizier to Haroun*

*. . . and a host of phantoms, genii, magicians,
sorcerers, philosophers, beasts, merchants,
hermits, enchanted rocks, and other beings*

BOOK I

The
REIGN
of
HAROUN
AL RASHID

WHERE IS
HAROUN AL RASHID?

Haroun Al Rashid—Ruler of the Faithful—had gathered from sorcerers all over the world many secrets of magic. One of

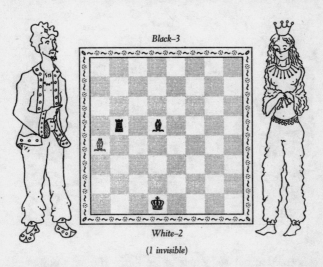

Black–3

White–2

(1 invisible)

his favorite tricks he had learned from a Chinese sorcerer (whose name, unfortunately, I cannot remember!). It was the art of invisibility. So here is Haroun standing in broad daylight, on one of the sixty-four squares of the enchanted chess kingdom. But nobody can see him, for the simple reason that he is invisible.

On what square does he stand?

INVISIBLE BUT NOT
INVINCIBLE!

Now the Black king, Kazir, has made himself invisible. Although it is during an important battle and Haroun is greatly

Black–2
(1 invisible)

White–6

concerned at not being able to see Kazir, Haroun's powers of reasoning are so remarkable that he is able to mate the Black monarch as unerringly as if he were in full view—and that in only one move!

What is the move?

4

HAROUN IN DISGUISE

As is well known, the Ruler of the Faithful would often go disguised among his people at night, to hear their reactions to what went on in the palace. So here is Haroun masquerading as

Black

White

some other piece—possibly an officer, possibly a pawn; possibly White and possibly Black. Which piece is he?

◈~◈~◈

Haroun learned many remarkable things that night. Since no one recognized him, his subjects spoke quite freely, with no idea that they were being overheard by the great Ruler of the Faithful himself. They spoke about all sorts of things. Many talked about matters pertaining to their daily lives, or about politics; some discoursed on the nature of fate and ultimate reality, others spoke of magic and the supernatural. But it was mainly people's reactions to the affairs of the palace that interested Haroun. Fortunately, everyone was in a rather festive mood that night and had only good things to say about the caliph.

Haroun was not at all the type who objected to being praised. Indeed, he so thoroughly enjoyed the episode that on the next night he decided to go forth again in disguise to hear some more good things said about himself. Only on this second night, he costumed himself differently than he had on the first. Where is he now?

Black

White

6

◈ 4 ◈

STORY OF THE
ENCHANTED ROCK

One day Haroun was wandering through the magic chess forest. It was a hot and sleepy day, and when he got to square g4 he decided to rest. He sat down with his back against a tree,

Black

White

and for a footstool used a nearby rock. To Haroun's utter astonishment, a voice came from the rock: "Please, sir, this is uncomfortable."

Haroun jumped to his feet in amazement and shock. "Was that *you* talking?" he asked the rock. "Yes," was the unbelievable reply.

"Who *are* you?" asked Haroun.

"That I cannot tell," replied the rock. "If I did, it would mean instant death for both of us."

"How can that be?" asked Haroun.

"Well, it is a sad story. Several hundred years ago I made a

7

clever move and gave offense to an evil chess genie. In wrath he turned me to stone and said: 'You shall remain in this condition forever, unless someone comes along who can correctly guess your identity. If the person makes a wrong guess, *he* will perish on the spot. And if you tell him who you are, *both of you* will die.' All I can reveal is that at the time of my transformation it was Black's move and the magic chess forest was just as it is at present."

Now, Haroun was unwilling to risk his life by *guessing*, but his great deductive powers soon enabled him to *know* the identity of the rock.

And so, dear reader, the problem is: What are the color and denomination of the unknown piece on g4? A second interesting problem arises with this situation: Can Black castle?

THE HIDDEN CASTLE

Hidden in the chess forest is a wonderful
enchanted castle, shining White. In winter,
when the leaves are bare, it is easily visi-

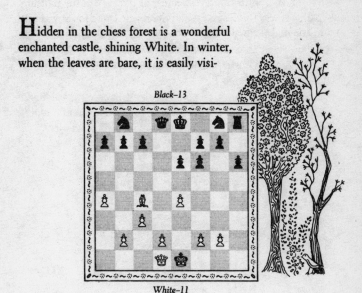

Black–13

White–11

(1 invisible)

ble. But now it is summer, and the castle is hidden by the deep
foliage—that's why the reader can't see it, although I assure
you it is there!

To find the castle, the reader must know the following fact
about the history of the game: None of the royalty has yet
moved or been under attack.

Where is the enchanted White castle?

◈ 6 ◈

A VITAL DECISION

One night, when things were slow at the palace, Haroun entertained the gathered company with an ancient account of how chess was

used for military purposes at the Battle of Saraca, a major engagement between Persians and Hindoos. This is the tale he told:

A small but important detachment of Persian forces was cut off from its fellows and was sure to perish unless reinforcements arrived. However, there was great risk involved in sending them, and if it was to be done, it was imperative that it should not be known to the enemy. The head of the Persian troops slipped a spy through to the isolated detachment with instructions to send back a coded message as to whether to send reinforcements or not.

The spy did as bidden, and when he arrived, came to an immediate decision about what should be done. He dispatched three homing pigeons, each carrying a separate message, none of which was intelligible without the others, but which together gave definite directions as to whether to send reinforcements or not. In this manner the message would be safe from enemy detection, unless, of course, all three pigeons should fall into the hands of the Hindoos.

It was exceedingly improbable, but as fate would have it, all three pigeons *did* fall into the hands of the Hindoos, and it was crucial that the messages be quickly deciphered.

The first pigeon carried a message that when deciphered read:

> White – No
> Black – Yes

The second pigeon carried the diagram on the previous page.

The third pigeon carried a note which when decoded read:

> Who moved last?

What is the solution?

MYSTERY OF THE
BURIED CASTLES

"Where are my two castles?" Haroun asked his vizier one day.

"They're not both yours," replied the impudent vizier.

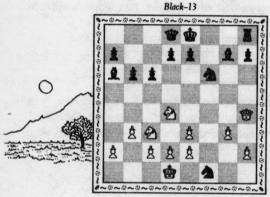

Black—13

White—12

"One of them belongs to Mrs. Haroun Al Rashid, you know."

"However that may be," replied Haroun, "I still want to know where they are."

"If you're referring to the two White castles, they were both captured during this game and buried on the spots where they were captured."

"But that is terrible!" said Haroun. "Some of my most valuable treasures are there—this very manuscript, for example! We must dig them up immediately!"

"Sixty-four squares is a lot of excavating, your majesty!"

"Well, by the beard of the Prophet, man, don't you recall *where* they were captured?"

"No," replied the vizier. "Nobody does. All that is recorded is that they were both captured on the same row."

"Not much help," replied the exasperated caliph.

"No," said the vizier.

"Well, *do* something!" roared the caliph.

What could he do? He consulted the chronicler, who, though he did not know where the White castles were buried, did recall that the Black queen's rook was captured by a pawn.

"This information may help," said Haroun. "I also recall, if I am not mistaken, that you did not participate in this game. Am I correct?"

"Yes, your majesty, I did not participate. I was sick the first day of battle, and had to be removed from the field. I was, as they say, given as odds of a bishop."

"Sick, or lazy?" roared Haroun.

"Sick," replied the vizier.

"Well now, let us not quibble; far bigger issues are at stake. *I must find my castles!*"

"You mean the *White* castles," corrected the vizier.

"Oh, hang it, man, will you be quiet! Let me think: Both my castles captured on the same row, Black queen's castle captured by a pawn, and king's bishop given as odds. Won't do; the problem is unsolvable!"

Just then, however, news came that Black was about to castle. "Praise be to Allah," cried the caliph, "the mystery is solved! Now I can find my treasures!"

Where are the buried castles?

CASE OF THE
DISPUTED CASTLE

On square a5 stands a valuable castle filled with wonderful treasures. But since the castle is very old, many of its stones

Black–13 or 14

White–13 or 14

have been replaced by stones of the opposite color, and the castle appears half black and half white. Because of the treasure, both sides claim the castle for their own.

Now, it is known that White has no promoted pieces on the field and that Black has just castled. Which side really owns the disputed castle?

MYSTERY OF THE
BLACK CASTLE

Near the center of the magic chess forest stands a glowering, grim, evil-looking Black castle. And if you think it looks black and

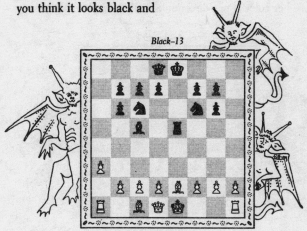

Black–13

White–13

wicked on the *outside*, you should see the *inside*! And if you think it *looks* vile, foul, and corrupt on the inside, you should know what *goes on* in the inside! Shall I tell you what goes on in the inside? All right, I'll tell you what goes on in the inside! The most vile, foul, evil, and wicked deeds—that's what goes on in the inside. In short, this castle is a Black den of iniquity. And this iniquity must stop! The owner of the castle is responsible for all that goes on and must be brought to justice.

White has given Black odds of a knight, and none of the royalty has yet moved. Does the Black castle belong to the Black king Kazir or his queen Medea?

STORY OF
THE VEILED QUEEN

On h5 stands a mysterious, beautiful queen, veiled from head to foot. Although every square inch of her skin is covered, and hence her color is not discernible, both kings are so intensely

Black–13 or 14

White–14 or 15

aroused by her highly provocative form that each uncompromisingly insists that she is his own. In fact, the kings have been driven to such insane desperation over the issue that the bloodiest war in Arabian history has been raging these five years! And, sad to relate, lust has led to even more outlandish battle practices than would desire for material conquest. Daily, hundreds of thousands of men eat each other alive, and perpetrate other atrocities I cannot even mention to my gentle western reader. Meanwhile the queen stands coyly by, never so much as dropping a hint as to her true identity. Of course she could be forcibly disrobed, but that would be completely contrary to Arabian custom.

One day, after five years of this hideous barbarism, a mediator arrives on the scene. He says to both kings: "Gentlemen, be reasonable!" Instantly the fighting stops. Both armies crowd around to hear the proposed solution. The mediator says: "The veiled queen is, I presume, original rather than promoted?"

"Of course," reply both kings. "We would not think of making such a fuss over a *promoted* queen."

"Very good," replies the mediator. "Now, are there any other recorded facts about this game?"

The only information that anyone remembers is that no pawn has ever captured more than one piece.

The mediator says: "Aha! Aha!" and thinks for a while. A bright smile suddenly beams across his beamish face. He then says: "And now, the crucial question! Was the missing queen captured on or off her own file?"

"Of what account is that?" ask both kings simultaneously.

"Of the utmost," is the quizzical reply.

Prove that the mediator is right.

≈ ≈ ≈ ≈

As explained in the solution, the mediator was indeed right! The color of the veiled queen depends absolutely on whether the other queen was captured on or off her own file.

As fate would have it, neither side *could* remember whether the missing queen was captured on or off her own file. Hence the battle was shortly resumed, and is raging till this very day—unless, of course, it has subsequently stopped.

17

BOOK II

TALES
of the
TREASURY

STORY OF THE
PURLOINED TREASURE

"Some of my treasure has been stolen again!" cried Haroun furiously one day. "Why do people so delight in stealing from me, even while I'm in the midst of battle?"

The vizier laughed.

"My dear sir," replied Haroun, "this is no laughing matter!"

"Did he take very much, your majesty?" inquired the vizier, putting on a graver countenance.

"No, it was only a piddling amount, but, hang it, man, it is the *principle* of the thing! I can hardly have everyone in my kingdom helping himself to my treasures whenever the inspiration comes over him!"

"Of course, your majesty. Now, let's get down to details. Have you any idea who the culprit is?"

"Yes, indeed; I have had one of my spies on the case for several weeks, and he has proved to my complete satisfaction that the miscreant is the White pawn originally from f2—your very own page, as it happens!"

The vizier flushed visibly and said, "Now, Haroun, I hope you don't think *I* had anything to do with this!"

"Hardly, my dear fellow," laughed Haroun. "I know you better than to accuse *you* of stealing such a small amount!"

Ignoring this, the vizier asked, "Why don't you have him arrested and tried?"

"That's the rub of it!" cried Haroun, "I'm not sure where he is! The pawn on g3 appears suspect, but there seems to be no way of telling whether he is your pawn from f2 or the rook's pawn from h2. When interrogated, he *claimed* to be from h2. Is that not crafty of him?"

"If guilty, yes; if innocent, he is simply telling the truth. Now, draw me a diagram of the battlefield, and let me see if we can figure out whether g3 is guilty or not." The Great Ruler of the Faithful then dutifully sketched the diagram that appears below.

After a few moments sunk in thought, the vizier said, "Your majesty is slipping! It is not all that difficult to tell whether g3 is guilty or not! Have him tried tonight, and I will prove to your majesty's satisfaction his innocence or guilt."

Is g3 guilty or innocent?

Black–10

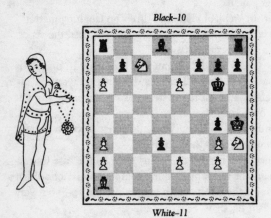

White–11

PURLOINED
TREASURE II

So the true miscreant got away, and news of the theft reached the Black forces. Barab the pawn—who in this game started out on b7—was encouraged by this report and thought, "If one

Black–15

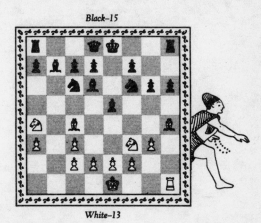

White–13

pawn can get away with robbery, why can't another?" So he stole a little from the Black treasury and ran away, planning to reach White's first rank and forever conceal his identity.

When Kazir heard of this, he was quite put out. He knew the position, but could not figure out whether Barab had actually made his goal or gotten captured along the way. A bright idea suddenly occurred to him. He called a truce with White and sent a quick message to Haroun: "Are you able to castle?" Haroun sent back a quick reply: "Yes." Kazir heaved a sigh of relief; this information provided the vital clue! Kazir was able to deduce whether Barab had made it or not, and if so whether he was still on the battlefield in a converted form. His analysis is perhaps one of the most profound in the entire chronicles of the Arabian Knights. What is it?

PURLOINED
TREASURE III

As brilliant as Kazir's analysis of Barab's movements was in the last episode, he was powerless to ascertain just which Black promoted officer he was. Thus Barab escaped the wheels

Black–15

White–14

of justice. In the present episode, Barab is less fortunate.

A new battle has begun. This time, Barab starts out on a7 and tries the same trick again—steals some treasure and rushes toward White's first rank with hopes of again concealing his identity by promoting. Now, one of Kazir's spies actually sees Barab crossing from White's second to first row, so it is known that he has promoted. Also, the spy reports, both kings have moved only once. From this information it is quite easy for Kazir to ascertain which of the promoted Black officers is Barab. Which one is he?

❖ 14 ❖

PURLOINED
TREASURE IV

When Gary, a White pawn, heard of Barab's unsuccessful escapade, he decided that he could do the job more subtly. And so he helped himself to some of Haroun's treasure, trotted

Black–14

White–13

off to the eighth rank, and promoted. Unfortunately for him, he was spied swimming across the river separating the seventh and eighth ranks. Also, it was known that the Black queen had been captured on her own row, the White queen's rook captured on its own file, that neither king had moved, and that Black could castle. These facts proved sufficient to locate him. Where is he?

26

TALE OF THE
WILY BISHOP

One day Haroun found an *enormous* sum missing from his treasury. "Hah!" cried Haroun. "No one could steal such a huge amount except my own bishop! By the beard of the

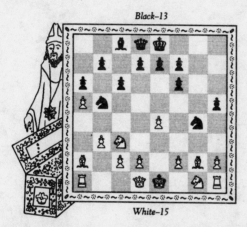

Black–13

White–15

Prophet, I will have him arrested within the hour! Here! Captain of the Guards!"

How great was Haroun's consternation when the captain of the guards returned dismayed after a brief scout and said, "O Ruler of the Faithful! There are *two* White bishops on the field, and both on white squares—a2 and g2. Which one do I arrest?"

You see, the crafty bishop had bribed an innocent White pawn to assume his likeness upon reaching the eighth row.

How Haroun stormed and raged! "Well now," he cried finally, "I will use my wits and *deduce* the whereabouts of the deuced bishop." He did this in a reasonably short time, sent the captain back onto the field, and the guilty bishop was brought back in chains *just* within the hour. Thus the Prophet's beard was saved.

Which is the original and which is the promoted bishop?

SECOND TALE OF THE
WILY BISHOP

Although it was a relatively simple matter to find the king's guilty bishop in the last episode, the queen's bishop thought he would try the same thing—with variations. So he also stole

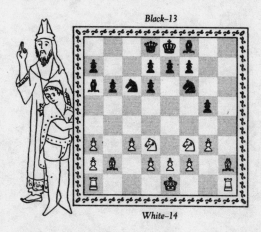

Black–13

White–14

some treasure and bribed a White pawn to promote to a bishop on black squares. But the position involved was much more difficult to analyze!

Indeed, from the position alone it was not possible to deduce which was the original White bishop. However, Haroun's historian had chronicled three facts about the game which were sufficient to convict the original bishop:

1. The rook from h8 was captured on its own square.
2. Neither king had yet moved.
3. There were no promoted Black pieces on the board.

Which is the original White bishop?

ARCHIE'S COUP

When Archie, the king's bishop
(and also his grand vizier), finished
his prison term for stealing the

Black–4

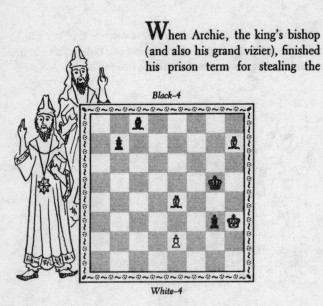

White–4

king's treasure, he decided that this time he really *would* out-
wit Haroun. So he helped himself to a good deal more of the
precious hoard and tried the same old trick again.

Where is the king's bishop *now*?

A PAIR
OF ODD-BALLS

The oddest robbery of all was committed by a White piece (or pawn) that did not even participate in the game! It was technically "given as odds." Which piece is it?

Black—11

White—11

For some curious reason or other, many of the Arabian Knights tales—and especially those relating to chess—have been told in more than one version. How these different versions arose would be an interesting subject for historians. The most authentic source of the tales is, of course, the *Book of Shah* (unfortunately, this remarkable work is exceedingly rare and appears to be as little known to Western readers as the somewhat comparable volume *Tellmenow Isitsoörnot* referred to by Edgar Allan Poe in his "A Thousand and Second Tale of Scheherazade"). The *Book of Shah* has been variously referred to as the *Book of Chess*, the *Book of Kings*, and sometimes as the *Book of Wisdom* (the latter because it contains not only historical episodes but many philosophical observations on life in general). But even the *Book of Shah* sometimes contains more than one version of the same story.

The present tale is a good case in point. According to an alternate telling, the situation is the one shown below.

White has similarly given Black odds of a piece or a pawn, and, in addition, there have been no promotions in this game.

What was given as odds?

Black–11

White–11

BOOK III

ARABIAN
KNIGHTS

CASE OF THE
LAZY KNIGHT

"**W**hy is my knight so lazy?" Haroun asked his vizier one fine day. "I understand he has moved only once this entire game!"

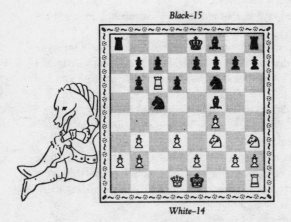

Black—15

White—14

"Why don't you have a little talk with him?" suggested the vizier.

"That's the trouble!" said the enraged caliph. "There are White knights on both f3 and h3, and I don't know which is mine and which is Amelia's! And both knights, of course, claim to be Amelia's! Those two are always covering up for each other!"

"Why don't you remove their armor?" asked the vizier. "Then, when you see their *faces*, you might have a better idea as to which is which."

"That's impossible," replied the caliph. "It happens to be illegal to disarmor a knight. The only way I could disarmor a knight would be if Kazir and I had a disarmoring conference, but so far none has ever been called."

"Hm," said the vizier gravely, "that makes it a bit difficult, doesn't it?"

"Yes," replied the caliph.

At this point the vizier had a bright idea. "Why don't we try to *deduce* which knight is which?"

"Not much chance," said Haroun sadly, "I don't remember much about the history of this game. All I know for certain is that *I* have not yet moved."

"Did you say your *knight* is lazy?" innocently inquired the vizier.

"That's enough out of you!" roared the caliph. "One more crack like that, and you go straight back to the torture rack! I still haven't forgotten that huge amount of treasure you last stole from me!"

"I only did it for *your* sake," protested the vizier.

"What in the name of the Prophet are you talking about?" asked the caliph.

"Haroun, old friend," said the vizier, "you know that I have more than enough money as it is; I would hardly steal from your sacred majesty's treasury just for the sake of adding to my own personal wealth."

"Then why on earth *did* you steal it?"

"Only to show the world your magnificent cleverness in apprehending me."

"What?" exclaimed the stunned caliph.

"Of course, your majesty," replied the vizier. "You recall how clever *I* was after the theft in getting a pawn to promote to my likeness, and how you were even more clever in deducing which was the promoted pawn and which was me?"

"Of course I recall," said the caliph, "and so what?"

"That's the whole point!" excitedly exclaimed the vizier. "If I hadn't committed this theft, you never would have had this marvelous opportunity to display to the world your magnificent skill as a retrograde analyst. Just think! Without this crime, you would not have gone down in history as the solver of the famous Mystery of the Promoted Bishop. Indeed, without this crime, and similar ones I plan to commit in the future, this whole book of 'Arabian Knights' might never have been written."

"Did I hear you say *similar crimes you will commit in the future?*"

"Oh yes, your majesty," said the vizier, "but all of them only for your sake, you know; all of them just to bring you honor and renown for your skill in catching me."

"A likely story!" exclaimed Haroun.

"As unlikely as can be," responded the vizier, "but nevertheless true."

"At any rate," said Haroun, "how does all this help us solve the present mystery of the knight?"

"Not at all, as far as I can see," replied the vizier.

"Then why are we wasting time talking about it?"

"I have no idea," replied the vizier. "It was you, not I, who brought the matter up."

"I only brought it up," said Haroun, "because of your innuendo about my having not moved and being lazy like the knight."

"Well, at least *that* was relevant to the present problem," said the vizier. "Anyhow, whatever your motive was in bringing up the matter of my theft, it was you and not I who brought it up. So don't blame *me* for wasting time!"

"Meanwhile we are wasting *more* time arguing about who is responsible for wasting time," said Haroun.

"Exactly," replied the vizier, "and don't blame me for that either."

"Then what should we do?" asked Haroun.

"I suggest we consult the chronicler," replied the vizier.

So the chronicler was brought in and consulted. Now, this chronicler was a most remarkably eccentric chronicler! It was not that he ever recorded false facts; everything he recorded was thoroughly reliable. Rather, his eccentricity lay in the utter haphazardness of the facts he recorded. He was totally unsystematic, and had no sense whatever of what is sometimes referred to as "discipline." Since he lived *when* he lived, he had never been exposed to the Protestant "work ethic," and accordingly enjoyed a very easygoing, leisurely, quiet, and dreamy life. Indeed, if he hadn't lived *where* he lived, he might have been described as a Taoist. He usually got up very late in the morning, and if the mood suited him, he would saunter forth and record any facts of the game in progress that might happen to tickle his fancy. And being by nature truly Taoist (and, accordingly, influenced very much by the principle of negativity and nonbeing), he had this extremely curious habit of recording things that *didn't* happen as not happening, rather than things that *did* happen as happening. For example, in this particular game, he recorded—of all things!—that the square d7 had never been occupied more than once nor traversed more than once. *Why* this particular fact should interest him is utterly beyond me! And yet, this very fact turned out to be a vital clue in ascertaining the identity of the lazy knight. Another equally important fact was that the Black queen's knight had moved only twice.

So then: (1) the White king has never moved; (2) the Black queen's knight has moved only twice; (3) the square d7 has never been occupied or traversed more than once.

The White king's knight has moved exactly once. Is he the knight on f3 or the knight on h3?

LAZY KNIGHT?

"That knight of mine!" Haroun fumed one day to his vizier. "This time he doesn't seem to have moved at all!"

"*Seems* not to, or is *known* not to?" inquired the vizier.

Black–14

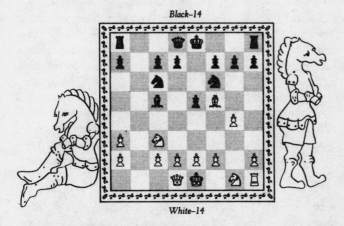

White–14

"Well, I don't *know* that he hasn't moved," admitted the caliph. "However, he is still on his home square."

"The phrase '*still* on his own square' appears to beg the question," responded the vizier.

"True," admitted Haroun. "Yet I don't *recall* his having moved."

"That's hardly evidence that he hasn't."

"True," said Haroun. "What do you suggest we do?"

"Draw me a diagram of the position," said the vizier.

"I have," said Haroun.

The vizier studied it for a while and asked: "Haroun, are there any promoted pieces on the board?"

"No," replied Haroun, "I do know that there are no promoted pieces on the board."

"Then your knight is not so lazy," said the vizier. "It is relatively easy to deduce that your knight has moved."

How did the vizier know?

WHICH
LAZY KNIGHT?

"Now I have another problem," said Haroun, a few days later. "The chronicler has reported to me that again one of the two White knights hasn't moved."

Black–10

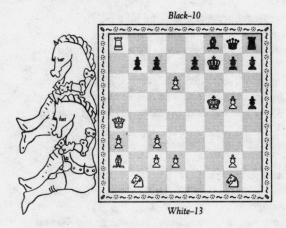

White–13

"Which one?" inquired the vizier.

"That's the trouble!" said the caliph, turning more and more livid with rage. "He is so damned absent-minded, he neglected to record which one—and he doesn't remember, either. He is the most eccentric chronicler I've ever had! I think one of these days I will have him executed!"

"Not a good idea, your majesty," replied the vizier. "Erratic as he is, he has an uncanny knack for recording just those odd little facts that are useful to us in our retrograde investigations."

"Which of the knights *is* the lazy one?" shouted the caliph, stomping all around the throne room.

"Calm down, your majesty," said the vizier. "The matter is hardly that important. All this fuss over a lazy knight!"

"Unimportant?" screamed the caliph. "Unimportant? Don't you realize that one of the knights is *lazy*?"

"So what?" asked the vizier. "What's so bad about being lazy?"

"What's so bad?" asked the caliph, unable to believe his ears. "What's so *bad*? You weren't being lazy when you stole my treasure!"

"That's hardly the argument to bring up *now*," laughed the vizier. "If anything, it is an argument in *support* of laziness. Wouldn't you rather I had been more lazy and *not* stolen your treasure?"

"Sheer sophistry," exploded the caliph. "Everyone knows it's wrong to be lazy. Haven't you heard of self-discipline? Haven't you heard of the evils of idleness? Haven't you read the Koran?"

"Yes," replied the vizier, rather sadly.

"Well?"

"Well?" replied the vizier.

"Now don't be impertinent!" screamed the caliph. *"Find me the lazy knight!"*

"Draw me a diagram," demanded the vizier, and Haroun produced the one on the previous page.

"I see which is the lazy knight," said the vizier.

Which one is it?

TALE OF A
MISCHIEVOUS KNIGHT

"These knights!" said Haroun. "Always giving me head-aches!"

"Another case of laziness?" asked the vizier a bit wearily.

Black–12 or 13

White–5 or 6

"No, not laziness this time, just silly playfulness."

"So what's wrong with a little playfulness?" asked the vizier.

"What's wrong?" shouted Haroun. "You weren't so playful that time you stole my treasure!"

"Why are you so illogical?" asked the vizier. "You always bring up the wrong argument—an argument that tends to establish the very opposite of what you are trying to prove."

"What do you mean?" asked Haroun.

"Well—and this is typical—here you are talking *against* playfulness, and you blame my stealing your treasure not on

43

my playfulness, but on my *lack* of playfulness. What kind of logic is that?"

"It is very bad to be too playful," said the caliph—completely ignoring the vizier. "There is serious work to be done, and one shouldn't remain a child forever."

"Oh, for heaven's sake!" said the vizier. "Let's not get into another silly philosophical discussion. Let's go immediately to the problem of your overly playful knight. Which knight is it, yours or Amelia's?"

"I don't know that it's either," replied Haroun. "It may be one of the Black knights."

"This is beginning to sound interesting," replied the vizier. "Tell me more."

"Well," said Haroun, "the situation is like this." And he showed the vizier the diagram that appears on the previous page. "On a1 stands a knight dressed in armor that is half black and half white; I can't tell his color. I know he is doing it just to bug me!"

"And you can't disarmor him because you haven't had a disarmoring conference. Right?"

"Right," said Haroun.

The vizier studied the situation for a few minutes. "Haroun," he asked, "whose move is it?"

"It's Black's move."

"Ah! Then I know the color of the knight."

What color is the knight?

THE KNIGHTS
WHO CHANGED ARMOR

"More mischief!" said Haroun. "This time two of the knights, one White and one Black, decided to change armor. Thus one of the knights who appears Black is really White,

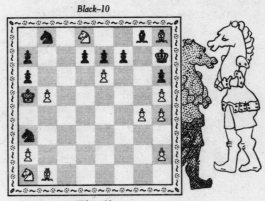

Black–10

White–12

and one who appears White is really Black. It's very confusing."

"How do you know this?" asked the vizier.

"The chronicler told me," replied Haroun.

"And, of course, he neglected to tell you *which* two knights are up to mischief."

"Of course," said Haroun. "You know our absent-minded chronicler!"

"Hm," said the vizier. "Let me study this situation seriously."

After a while the vizier said, "Haroun, I'm really glad this situation came up. The solution is one of the most beautiful I have ever seen. I know which two knights have changed armor."

Which are they?

AN ANCIENT PUZZLE

"Haroun," said his vizier the next day, "the last situation reminds me a bit of a very ancient chess puzzle I came across some time ago.

Black—12

White—12

"It is given that in this game Black gave White odds of the queen's bishop, and that no knight was ever captured by a pawn, and that Black can castle. On c8 and h1 stand two knights of opposite color. The problem is, which is which?

"This problem," continued the vizier, "has an interesting twin. Move the Black bishop from h2 to a2, and change the given conditions as follows: No knight or pawn or promoted piece has been captured by a pawn, nor has any pawn capture been made on either the first or eighth row. Again Black can castle. No odds were given in this game. Now which knight is which?"

THE INVISIBLE KNIGHT

"Now my troubles are serious," said Haroun one day. "This time it is not just a mischievous little prank—but one of my most precious treasures has been stolen."

"What treasure is that?" asked the vizier eagerly.

"Why, my powder of invisibility! I need it!"

"By powder of invisibility, do you mean powder that is invisible?"

"Of course not, you idiot! Invisibility powder is not invisible; it is not to be confused with *invisible* powder. The former is definitely visible; it has a pale bluish-white color, in fact."

"Then why is it called invisibility powder?" asked the vizier.

"Because when you take it—when you swallow it—it makes you invisible."

"But the powder itself is visible?"

"Yes, I just told you that!"

"Even after you take it?" inquired the vizier.

"No, not after you take it! It then becomes invisible like the rest of you."

"But doesn't that contradict your original statement that invisibility powder is visible?"

"Not really," said Haroun, getting a little irritated. "It becomes part of your system, and ceases to be powder."

"But how can anything cease to be what it is?"

"That's an interesting philosophical problem, but I don't want to get sidetracked! I want to locate my powder of invisibility."

"Who has stolen it?" asked the vizier.

"One of the White knights."

"So why don't you have him apprehended?"

"Because, you numbskull, he is *invisible!* Do you think he would steal my invisibility powder without using it himself? What better way can he escape apprehension than by being invisible?"

The vizier thought about this for several minutes and became increasingly puzzled. "Let me get this straight," he said. "He took the powder for the purpose of becoming invisible."

"Naturally," said the caliph.

"And he wants to be invisible so as to avoid being apprehended."

"Of course," said Haroun.

"Apprehended for the crime of stealing the powder."

"Right," said Haroun.

The vizier was more puzzled than ever. This seemed to him to be an unnecessarily roundabout way of avoiding apprehension. Oh, well, he thought, people certainly are oddly paradoxical at times!

"Well, don't just sit there *thinking*," said Haroun. "*Do* something. Help me recover my lost powder!"

"Maybe we should ask the chronicler," suggested the vizier. "Perhaps he knows."

"Excellent idea," replied Haroun. So he dispatched a message to the chronicler: "Have you seen my invisible knight?" The chronicler promptly replied: "Of course not! What a stupid question!" Haroun sent back: "I mean to say, do you *know* where my invisible knight is?" The chronicler replied: "No, I have no idea where your invisible knight is, and I couldn't care less!"

This last message—and particularly the last clause of the last message—seemed to rub Haroun the wrong way; it somehow savored of the impertinent. "Have the wretch brought here in chains!" Haroun demanded.

And so the chronicler was brought to the palace in chains, where Haroun could comfortably interrogate him. "I want to get to the bottom of this," roared Haroun. "If the knight is not found within twenty-four hours, your own life shall be forfeit! Now, you say you have not seen him anywhere?"

"How could I possibly *see* an invisible knight?" the chronicler protested.

"I mean, you have no idea of his whereabouts?"

"Of course not! If I had, wouldn't I have told you by now?"

"That's better!" said Haroun. "I like it when people answer me a little more courteously. Now here is the present position."

Black–13

White–14

(1 *invisible*)

Thereupon he produced this diagram. "What," he asked the chronicler, "do you remember about the history of the game?"

"Not much," said the chronicler rather sadly. "I do know that the Black king and queen have not yet moved, nor been under attack—if that is of any help."

"It may be," said Haroun. "It is also true that neither Amelia nor I have yet moved. Anything else?"

"I recorded the fact that the Black rook on a5 has moved only once."

"Strange thing to record," said Haroun. "However, it may be helpful. Who knows? Anything else?"

"I also recorded Black's first move of the game—it was with the pawn on c5, which came from c7. Does that help?"

The vizier wrote down the three facts: (1) None of the royalty has moved, and the Black king and queen have never been under attack; (2) The Black rook on a5 has moved (only) once. (3) Black's first move was the pawn from c7 to c5. He studied the situation for several minutes. Suddenly he brightened and said: "Haroun, you can set the chronicler free! I now know where the invisible White knight must be."

Where is he?

WHICH IS
THE GUILTY KNIGHT?

"**A**rchibald," the caliph said one day to his vizier, "we have
a serious problem on our hands. It is really
a case for a detective. I wish Sherlock

Black–15

White–14

Holmes had been born! It would make my life so much
easier!"

"What is the problem?" asked the vizier.

"Well, briefly, the situation is this: A rather grave crime has
been committed by one of the White knights. It is known that
he committed it on his third move. It is also known that the
other knight has moved exactly twice. That's all that's known
about the knights. So how do I tell which is the guilty knight?"

"Interesting," said the vizier as he studied the position carefully. "It certainly seems that *either* knight could have come
from b1 in two moves, and the other from g1 in three."

"Exactly," said Haroun.

The vizier studied the position some more. Suddenly he had a bright idea: "Tell me, Haroun, do you happen to know whether there are any promoted pieces on the board?"

"No, there are not," replied Haroun. "This I do know, but I hardly see that that is relevant!"

"Most relevant, your majesty; now I know which is the guilty knight!"

Which one is it?

HAROUN'S VIZIER SOLVES
A MURDER MYSTERY

Haroun one day called his vizier for an emergency meeting. "The case is urgent," he said. "Olaf's life is at stake; there is not a moment to be lost!"

Black–12

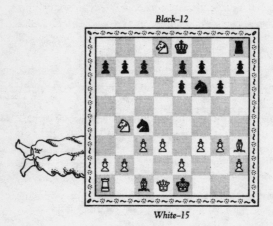

White–15

"How come you are so anxious to save Olaf's life after he played all those pranks on you?" inquired the vizier.

"Oh, come on now!" said Haroun. "Fun is fun, but this is serious!"

"I thought you didn't approve of fun; I thought you believed in hard work."

"Now look," said Haroun, "of course I don't approve of fun! But I have already punished Olaf for his mischievous pranks. I may be a moralist, but I'm not a fanatic! There's a limit, you know."

"True," said the vizier. "So, what's the problem, and what can we do?"

"Well," said Haroun, "Kazir has charged Olaf with murder. The situation is this: Medea has been taken, and Kazir wishes to have her assassin executed. It so happens that Olaf stands on her square.

"From this, Kazir concludes that Olaf is guilty."

"The fact that Olaf stands on her square is hardly proof!" exclaimed the vizier.

"Of course not," replied Haroun, "you know that and I know that. But Kazir's ideas of justice are very different from ours! To him, a man is guilty unless proved innocent. So he has sent me an ultimatum that if within twenty-four hours I cannot prove that Olaf is innocent, then Olaf must die."

"Well, well," said the vizier, "I'll do what I can. Tell me, have you moved in the course of this game?"

"No," said Haroun.

"That may help, but it's not enough. I suggest you arrange with Kazir to hold a trial this very night. I will act as Olaf's defense counsel."

So a trial was held that night. The only evidence the prosecutor could bring forth was that Olaf now stood on the Black queen's square. Admittedly, this evidence was only circumstantial, but it was—by Kazir's law—to be held as final, unless controverted by evidence that was absolutely conclusive. After the prosecutor had completed his presentation, counsel for the defense was allowed to speak.

"Well, your majesty," said Haroun's vizier to Kazir in a *very* respectful tone, "has his august highness yet moved in this game?"

"No," answered Kazir, "I have not yet moved. What of it?"

"Well," replied the vizier, "my caliph has also testified that he has not moved. Will the court accept his testimony as evidence?"

"Only if sworn on the holy Koran," replied Kazir.

Haroun swore on the holy Koran that he had not moved in this game.

"Now that it has been established that neither caliph has moved," said the vizier, "I can prove to the court's satisfaction that Olaf the knight cannot possibly have been the assassin of the Black queen."

He did this, and Kazir was forced to exonerate the knight. How did the vizier prove it?

BOOK IV

TALES
of the
PALACE

MYSTERY OF THE SPY

"Could you tell me a story?" Haroun asked his vizier one sultry day. "Things have been rather slow around here."

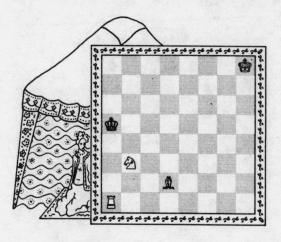

"What kind of story would you like, your majesty?"

"Well, do you know any spy thrillers?"

"Hm," said the vizier, "let me think! Why yes, your majesty, I know a spy mystery. I don't know if you would call it a thriller exactly, but it contains an interesting chess puzzle."

"Show it to me," cried the caliph eagerly.

"Well, your majesty, the situation is this." And the vizier set up the chessboard as shown above.

"In this game, one of the pieces is a spy in disguise. Which piece is disguised, and what is it really?"

SECOND MYSTERY
OF THE SPY

"That was a dirty trick," said Haroun, after the vizier showed him the solution of the preceding puzzle. "Show me

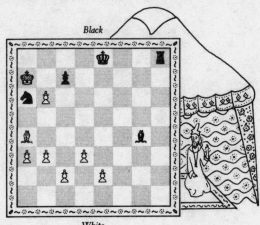

Black

White

another which is more straightforward; one that is honest."

"I think that was honest," replied the vizier. "However, this one may be more to your liking.

"One of the Black officers is a White spy in disguise. What is it really?

"A second question: Can Black castle in the future?"

STORY OF THE
MASTER SPY

"That's better," said Haroun. "I liked that problem. The solution was simple and pleasant. Know any more?"

"The best spy story I know," replied the vizier, "and indeed

Black–13 or 14

White–11 or 12

it might be called a 'master' spy story, is remarkable for the subtlety of its analysis. It is the most wonderful spy thriller in the annals of the Arabian Knights."

"That last statement," said Haroun, "strikes me as a bit peculiar, but go on."

"Once upon a time," said the vizier, "there was a *master* spy. He (or she) decided to disguise himself (or herself) so cleverly that it would be extremely difficult to detect him (or her). All he (or she) did was change color; he (or she) remained the same denomination.

"Now, all that is given is that the master spy is not a pawn. As I said, he (or she) is the wrong color but correct denomination. Which piece is it?"

TRIAL OF THE
BISHOP

"Excellent!" said Haroun. "That was indeed a master spy story. The spy certainly turned out to be the *least* likely piece of all! Now, do you know any of what they call 'detective' stories?"

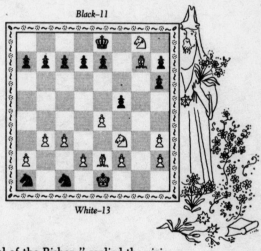

Black–11

White–13

"The Trial of the Bishop," replied the vizier.

"And what is the trial of the bishop?"

"Well, your majesty, the situation is this: One day a significant crime was committed somewhere on the eighth row. There was some evidence pointing to the guilt of the White queen's bishop.

" 'Impossible,' cried the bishop during the trial, 'it couldn't have been *me!* Why, I have never even *been* to the eighth row!'

"If the bishop hadn't said this, he might have gone free. But when he did say it, the White king—who was the judge—knew he was lying. He deduced this because it was already chronicled that it was now White's move, and that neither king had yet moved. Having discovered this lie, the White king investigated more closely, and found enough evidence to convict the bishop.

"How was it known that the bishop was lying?"

MYSTERY OF THE
LOST PAWN

"Nice story," said Haroun. "Tell me another detective story."

"Do you know the Mystery of the Lost Pawn?" asked the vizier.

Black–13

White–14

"No, what is the mystery of the lost pawn?"

"Well," said the vizier, "once upon a time there was a little Black pawn which started out on g7—"

"Sounds like Barab," interrupted the caliph.

"It could be Barab," replied the vizier. "Anyway, his father—the Black king—was worried about him, and wanted to know his whereabouts. If he was on the field, either in his original or promoted form, where was he? If he was captured, either in his original or promoted form, where was he captured? What *had* happened to him?

"His father was quite desperate, and finally had to hire what is called a 'detective' to find him out. The detective ascertained that the White royalty had never moved or been under attack. This proved sufficient to deduce the fate of the pawn.

"What happened to the pawn?"

THE LADY
AND THE KNIGHT

"Today I am in a romantic mood," said Haroun. "Do you know any love stories?"

"I know one," replied the vizier. "I learned it from Amelia—that is, Mrs. Haroun Al Rashid, queen of the great Haroun Al Rashid, Ruler of the Faithful."

"You *did?*" asked Haroun suspiciously. "When was this?"

"Now, now, your majesty," retorted the vizier, a bit flustered, "nothing to worry about! My relationship with Mrs. Haroun Al Rashid is a very beautiful one—but purely Platonic, you know."

"I *don't* know!" cried the caliph. "Moreover, I distrust anything Platonic. I happen to be an Aristotelian!"

"The story is a very beautiful one," said the vizier, hoping to divert the caliph from this particular line of inquiry.

"What is the title called?" asked Haroun.

"It's not called anything."

"The story has no name?"

"Oh, the *story* has a name, but not the title. Haven't you read Lewis Carroll?"

"Who is Lewis Carroll?" asked Haroun.

"Oh, never mind!" said the vizier.

"At any rate," said Haroun, "what is the title?"

"The story has two titles," replied the vizier. "One is *The Lady and the Knight* and the other is *Romance of the Lady and the Knight.* Which story would you like to hear?"

"I don't understand," replied Haroun. "I thought the two stories were the same."

"They are," replied the vizier, "and that should simplify your choice enormously!"

Haroun was decidedly puzzled by all this, and he spent several minutes trying to figure out what in the world was going on.

"The story is very beautiful and very touching," said the vizier with a deep sigh. "Every time I think of it, either it brings tears to my eyes, or else . . ."

"Or else what?" asked Haroun.

"Or else it doesn't," replied the vizier, in Lewis Carroll fashion.

"Well now, what is the story?" asked the caliph.

"The story is very beautiful and touching," replied the vizier.

"Yes, yes!" cried the irritated caliph. "You said that before. But what is the story *about*?"

"The story is about a knight who was in love with a lady," replied the vizier. There was a long silence.

"Is that all?" asked the caliph.

"No, no, your majesty! You see, the lady was a princess. . . ." More silence.

"So then what happened?" asked the caliph.

"Well, the princess was locked up in the castle of the Black king. It was imperative that she be rescued immediately! The only trouble was that the knight did not know which of the Black castles was the king's and which the queen's. He did remember that none of the royalty had yet moved. Still, he could not figure out which castle he should go to. Don't you find this story beautiful and touching?"

"Not a bit!" said the caliph. "What's so beautiful and touching?"

"Can't you see the *pathos* of the situation?" pleaded the vizier. "Here is the knight so in love with the princess, wanting so desperately to rescue her, but not knowing where to go. Doesn't this bring tears to your eyes?"

"Not a one," roared the caliph. "The story is not over, I hope?"

"No," said the vizier. "Fortunately it has a happy ending. As the knight stood pondering what he should do, his squire suddenly brought news that White was about to castle. The knight leaped with joy! Now he knew where he should go! Where?"

STORY OF THE
MAGIC CARPET

"I guess I'm not too fond of love stories," said Haroun with a yawn. "What about a fairy tale? Do you know any fairy tales?"

"Do you by any chance know the Story of the Magic Car-

Black–13

White–14

pet?" inquired the vizier, with a certain weariness in his tone.

"I think so," said Haroun thoughtfully. "I believe I heard it when I was a boy. But I don't remember it too well. What is the story of the magic carpet?"

"Well," said the vizier, "the version I'm about to tell you is not, perhaps, the most *authentic* one; still, it is of particular interest from a chess point of view. Very briefly, the story is this," and he rearranged the chessboard once again.

"Both sides can castle. The White queen has never been under attack, and the Black queen was under attack only once.

"One of the White officers now on the board has had access to a magic carpet and has been illegally flying around from square to square. Which officer is he, and on which square did he acquire the magic carpet?"

THE PHANTOM BISHOP

"**W**hy don't we hold a storytelling contest?" said Haroun one day to the vizier. "Perhaps we will attract some good talent to the court and learn some new stories and chess puzzles."

Black–16

White–13 or 14

"Good idea," replied the vizier.
So Haroun placed a poster in the marketplace:

> CHESS-STORY CONTEST TO BE HELD AT
> THE PALACE. GRAND PRIZE TO WINNER;
> LOSERS GET EXECUTED.

"Are you really going to execute the losers?" asked the vizier a bit anxiously.

"Of course not," laughed the caliph, "but it is good to keep them on their toes, you know."

"Don't you think that remark about execution might discourage some of them?" asked the vizier.

"Not the valiant ones," replied Haroun.

Well, the gala occasion arrived! But only four storytellers were sufficiently valiant to show up.

"Well, well!" said Haroun. "I see we have a small but select company. Let the stories begin."

The first contestant said: "I know a story about a phantom bishop:

"The White queen's bishop is a phantom bishop, and hence invisible. It may be on the board or off the board. If on the board, where is it? If off the board, where was it captured?"

PHANTOM BISHOP II

"I know another story about a phantom bishop," said the second contestant:

"The phantom bishop—again the White queen's—was not

Black–15

White–13 or 14

captured on its own square. Where is it, or where was it captured?"

TWO
PHANTOM BISHOPS

"I know a story about two phantom bishops," said the third contestant:

"The White phantom bishop—which is not necessarily the

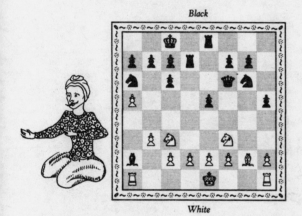

Black

White

queen's—is now on the board. The same piece was captured by the pawn on c6 as in the last two stories. Also, White can castle. Where is the White phantom bishop?

"The Black king's bishop has also become a phantom. Where is it, or where was it captured?"

75

BEST OF THE
PHANTOM BISHOPS

"**M**y story," said the fourth contestant, "involves only one phantom bishop. But it is quite remarkable:

"White gave Black odds of a knight. He has not yet moved

Black

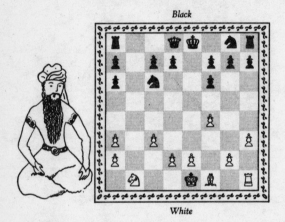

White

his king. Not more than one piece was captured on its own square, and that piece was not the first to fall.

"The Black queen's bishop is a phantom. Is it on the board or off the board?"

There was a unanimous judgment that the last story was the best. The unanimity was not surprising, considering that Haroun Al Rashid was the only judge.

BOOK V

TALES
of
MAGIC
and the
SUPERNATURAL

STORY OF THE
GENIE

"I must be going crazy," Haroun said one day to his vizier. "I have made a tour of the entire battle-

Black–13

White–15

(1 genie)

field—all sixty-four squares. There are three White knights on the field and eight White pawns. This is utterly impossible!"

"Hm," said the vizier, "one of the three knights must be a genie."

"A what?"

"A genie."

"What on earth is a genie?" asked Haroun.

"It is questionable whether genii are 'on earth,' " laughed the vizier. "A genie is a purely spiritual being that assumes—or rather simulates—material forms. Something like a mirage."

"A genie is a sort of phantom?" asked Haroun.

"No, no! Not at all like a phantom—rather it is exactly the opposite. A phantom is a piece that is really there but invisible; a genie is a piece that though visible is not really there."

"How can it be visible if it is not really there?" inquired the prosaic caliph.

"I told you, it is like a mirage. The genie, which is pure spirit, *appears* to have material form, but doesn't really."

The caliph thought for a while. "It gives me a rather creepy feeling," he said.

"Me too," replied the vizier.

The caliph thought some more. "It's really sort of spooky!"

"Yes," responded the vizier.

The caliph thought still more. "So you say one of the White knights is a phantom?"

"Not a *phantom*," replied the vizier, a bit irritated. "A genie. If it were a phantom, you couldn't see it. Please remember, you are not failing to see something which is there; you are seeing something which is not there. So it is a genie and not a phantom."

"The whole thing is a bit eerie," said the caliph.

"Yes," nodded the vizier wearily.

"So one of the White knights is a genie?" said the caliph.

"Yes, yes, *yes!*"

"How do you know it is a genie?" asked Haroun.

"Because that is the only rational scientific explanation for an otherwise impossible situation. As you say, there are eight White pawns on the board, so you can't have a promoted White knight. Therefore one of the three White knights is unreal."

Haroun thought about this. Suddenly he had a bright idea: "How do I know it is not one of the eight White pawns that is a genie?"

"Impossible," replied the vizier. "Genii are very snobbish

and elitist and would never condescend to materialize as lowly pawns."

"Oh," said Haroun. "I still don't like this whole business. I don't like having genii on my land."

"Me neither," said the vizier.

"Is there no way to dissolve a genie?" asked Haroun.

"Yes," the vizier replied, "by using a substance known as Genie-Dissolving Powder. If it is thrown into the face of the genie, the genie will silently and uncomplainingly dissolve. But if it is thrown into the face of a real being, the being will die instantly."

"That's not much help," replied Haroun. "We still don't know which of the three knights is the genie!"

"That is precisely the problem," said the vizier.

"You mean we have to *deduce* which is the genie-knight?"

"If you want to get rid of it, yes."

"How do we deduce it?" asked Haroun.

"Do you happen to know if Black can castle?" said the vizier.

"Yes," said Haroun, "Black can castle."

"Ah," said the vizier sadly, "if only you could castle in the future, that would solve the problem. But I guess you can't."

"Why do you say that?" asked Haroun in astonishment.

"Because you said you had made a tour of your entire kingdom. You must have moved to do *that*."

"Oh, no," said Haroun, "I did not make the tour *personally*; that would have been far too arduous! I sent my scouts."

"Then you *haven't* moved?" asked the vizier.

"Not once," replied Haroun.

"Praise be to Allah," said the vizier, "I now know which knight is the genie."

Which one is it?

SECOND STORY
OF THE GENIE

"That was a magnificent demonstration," said Haroun to his vizier, who had just shown him which knight was the genie. "Indeed," added the caliph magnanimously, "for this invaluable service, I shall buy you a new pair of shoes."

"Thank you," replied the vizier none too enthusiastically.

"And now," said Haroun, "give me the powder so I can dissolve the genie."

"I have no such powder," replied the vizier.

"What?" cried the caliph, turning purple. "You mean you have been deceiving me all this while?"

"I have not deceived you," pleaded the vizier, turning not purple but somewhat pale. "I never claimed to have any such powder."

"Why," roared the caliph, "I distinctly remember asking you whether there was a way to dissolve a genie, and you said, 'Yes, by using Genie-Dissolving Powder.'"

"True, I said that, and what I said is true. The one and only way of dissolving a genie is with Genie-Dissolving Powder. But this does not imply that I personally have any such powder."

"Miscreant," stormed the caliph, "you don't get your pair of shoes after all! How frustrating this all is—we know which knight the genie is, but we can't dissolve it—all for the want of the powder! What do you suggest we do?"

"Try to find some Genie-Dissolving Powder," suggested the vizier.

"Where can it be found?"

"I haven't the slightest idea," replied the vizier.

Well, they went together to the marketplace and first tried a general store. "Do you happen to have any Genie-Dissolving Powder?" Haroun asked the proprietor. "No," was the reply, "I have no Genie-Dissolving Powder." "Execute the wretch!" shouted Haroun. They went to another shop: "Do you have any Genie-Dissolving Powder?" asked Haroun. "Genie-Dissolving Powder? No, I have never heard of it." "Execute the wretch!" said Haroun. So they went from one shop to another—several thousand, all told—but none of them could supply Genie-Dissolving Powder. Most of the shop owners had not even heard of the substance, and those who had had none in stock. "Execute them all!" shouted Haroun. So all several thousand were led to the dungeons to await execution. Happily, Haroun rescinded all their sentences several years later and the world's shopping industry returned to normal.

"What do we do next?" asked Haroun.

"I guess we must go forth into the world and try to find Genie-Dissolving Powder."

"But I am at war with Kazir," replied Haroun.

"Perhaps he will agree to a truce until we return," suggested the vizier.

Fortunately, Kazir had just rediscovered Medea's charms and was pleased with the idea of a temporary truce so that he could spend more time with her. So Haroun and his grand vizier went forth into the world in search of Genie-Dissolving Powder. "Which way do we go? Where do we *look?*" asked Haroun.

"I suggest that we just wander aimlessly," replied the vizier. "It says in an ancient book of Chinese wisdom that sometimes the best way to attain an end is just to forget about it, to 'succeed by virtue of not striving.' Perhaps we might stumble upon some Genie-Dissolving Powder."

"Fat chance!" grumbled the caliph. Still, he had no better plan, so he wandered aimlessly with his vizier. Weeks went by, then months; still no Genie-Dissolving Powder. One day, weak and weary, about to give up, they came across the hut of a Chinese hermit. After he had refreshed them with delicious though simple food and cinnamon wine, he said: "Gentlemen, you look sad. What is your story?"

The caliph told him the whole business from the beginning—how there was a genie on the field who should be dissolved; that they knew where the genie was, but were powerless to act without the powder.

"How providential," replied the hermit. "I think I can help you."

"Really?" said Haroun, jumping up with joy. "You have Genie-Dissolving Powder? Where is it?"

"No," replied the sage. "Unfortunately, I do not have any such powder myself. But let me send you to a brother hermit who lives only a few thousand miles from here. I'm sure he is just the man you want."

They very carefully wrote down the address of the brother hermit, and the next morning left to pay him a visit. Several months later, weaker and wearier than ever, they found the brother hermit, who refreshed them with delicious though simple food and cinnamon wine.

"Now to business," demanded Haroun rather rudely. "I understand you have some Genie-Dissolving Powder."

"Praise be to Providence!" replied the hermit.

"What? You really have the powder?" exclaimed Haroun, jumping up eagerly. "Where is it?"

"Sit down," said the hermit, "and take it easy! No, I don't have the powder, though I can help you get it. My story is as follows.

"Seven years ago I was a merchant in good circumstances. One day at a bazaar I met a strange individual who offered to sell me a map for finding Genie-Dissolving Powder—the last in the world. 'Now, what would I want with Genie-Dissolving Powder?' I exclaimed incredulously. 'I have no genie I wish to dissolve.' 'Ah,' exclaimed the strange individual, 'purchase it anyway. I am a prophet, and can assure you that if you purchase it now, within seven years a mighty monarch will purchase it back from you at seven times the cost.'

"The price of the map turned out to be just about equal to all my worldly possessions! And I can tell you the whole idea struck me as preposterous. Yet some power stronger than my better judgment forced me to make this weird purchase. So I converted all my goods to gold, turned it over to this strange individual, and received the map. Utterly penniless, I have lived these seven years as a hermit. Now you have

Black–15

White–14

come asking about Genie-Dissolving Powder. Do you know a mighty monarch who might wish to purchase the map from me?"

"I am the monarch," said Haroun, rising. "I am Haroun Al Rashid, Ruler of the Faithful, and I will indeed pay you what you ask, if the map really turns out to lead us to the Genie-Dissolving Powder."

So Haroun and his vizier returned home with the map, amid the great rejoicings of the people. In Haroun's private study, they carefully examined the map. It was carefully laid out under the diagram of an ancient game of chess. Beneath the diagram was the legend:

> None of the royalty has yet moved. The Genie-Dissolving Powder lies buried with the White king's castle.

The message seemed clear enough, if only they could ascertain the square on which the White king's castle was captured. They did this without too much trouble. Can you?

Immediately they organized an excavating expedition, excavated on the right square, and, sure enough, uncovered an ancient White castle. Among the ruins, they found a container of Genie-Dissolving Powder. Then they went straight to the square on which the genie-knight stood and threw the powder in its face, and the genie silently and uncomplainingly dissolved away.

41

STORY OF THE
INCONSPICUOUS GENIE

Many, many years later, when Haroun and his vizier were old and white, they were reminiscing one day about their past adventures.

Black–13 or 14

White–13 or 14

"Remember that genie-knight and all the trouble we had getting rid of it?" asked Haroun.

"Oh yes," replied the vizier.

"Tell me," asked Haroun, "do genii always take the form of knights?"

"Oh, not at all! They can assume any form, except pawns."

"Have there been other cases of the appearances of genii recorded in history?"

"Oh yes," replied the vizier. "About two hundred years ago there was the situation you see here. It is given that none of the royalty has moved or been under attack. One of the officers is a genie and does not belong on the board at all. Which?"

✿~✿~✿

"About fifty years later," continued the vizier, "the genie reappeared in the following situation.

"This time, neither king has moved, and neither queen has been under attack. Where is the genie now?"

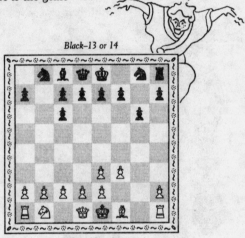

Black–13 or 14

White–13 or 14

42

ANOTHER TALE
OF A GENIE

One day Haroun received a message from the chronicler: "I saw a White bishop suddenly materialize out of thin air.

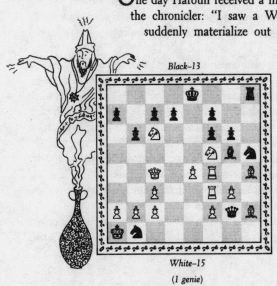

Black–13

White–15

(1 genie)

It must be a genie! I thought you would want to know."

"On what square was it?" Haroun wrote back.

"On a black square," was the reply.

"Yes, yes, I know that!" wrote back Haroun, "but on *which* black square?"

"That I don't remember," was the answer. "You know how absent-minded I am!"

"That chronicler will be the death of me!" cried Haroun to

his vizier. "One of the bishops on h2 and h4 is a genie; the chronicler *saw* him appear but has forgotten where!"

"Draw me a diagram," replied the vizier.

"I already have! The important thing is, we don't have any more dissolving powder!" cried the caliph helplessly.

"Forget your dissolving powder!" replied the vizier. "There is a much more important issue at stake—an issue of military significance! If one of the bishops is the genie, we can win the game in two moves! If the other one is, the case is hopeless!"

"Is there any way to decide which one is the genie?" asked Haroun.

"Not from the position," replied the vizier. "However, I have just got hold of a remarkable gadget known as a genie-detector—it was invented by that hermit who gave us the map. We will test each of the bishops separately to ascertain which one is the genie. If the one I hope is the genie *is* the genie, then we can move accordingly and win in two."

They proceeded to test the two bishops. Fortunately for the history of the kingdom of Haroun Al Rashid, the one that the vizier hoped was the genie was in fact the genie. Hence White won the game in two moves.

The problem then is this: Which of the two bishops on h2 and h4 must be unreal for White to have a certain win in two moves?

◈ 43 ◈

STORY OF THE
TRANSFORMED BISHOP

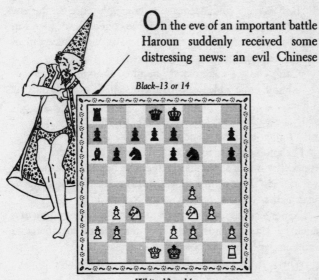

On the eve of an important battle Haroun suddenly received some distressing news: an evil Chinese

Black–13 or 14

White–13 or 14

magician had turned one of the White bishops into a beast (a horse/knight).

So one of the horse/knights on c3, c6, f6, or f3 is really a White bishop magically transformed. Haroun, of course, wants to rescue the bishop. However, he does not have his vizier around to consult with, since the vizier either has been captured or is one of the four beasts. So what can Haroun do?

Well, for once, Queen Medea—who herself is well versed in the arts of magic—has offered to come to Haroun's help. "I have here," she said, "a magic potion that, if fed to the correct beast, will immediately cause it to resume its former shape."

"And if it is the incorrect one?" asked Haroun.

"In that case, nothing will happen at all."

"Well then," said Haroun, "our course of action is logical and obvious. I have no idea *which* of the four beasts is the correct one, so we will simply feed the potion to each of them until the right one is found."

"Impossible," said Medea. "I have only enough for one dose. So if you don't know which is the correct beast, then I'm afraid I can't help you—or rather, I have one chance out of four of helping you."

"Can't you get some more?" inquired Haroun.

"Almost impossible," replied the queen. "I got this only by a fluke."

"Where did you get it?" inquired Haroun.

"From a Japanese magician," replied Medea.

The problem seemed pretty hopeless, but Medea volunteered the information that neither she nor her husband had yet moved. "That's interesting," said Haroun. "It also happens that neither my wife nor I have yet moved."

So, none of the royalty has yet moved. Was this information enough? Haroun and Medea put their heads together and came up with the correct solution.

Which of the four horses is really a White bishop?

STORY OF THE
ENCHANTED HORSE

Haroun Al Rashid was found one night in Medea's private
chamber. This, of course, Kazir could not tol-
erate. True, they were not found in a com-

Black—15

White—14

promising situation; nevertheless, the fact that he was found
alone with her at all was definite grounds for suspicion.

So there was a trial. The entire court was assembled. Haroun
was made to sit at a table upon which were two folded pieces of
paper.

"We shall let Fate decide your guilt," said Kazir. "On one of
the slips of paper is drawn a heart; on the other a dagger. You
will choose one. If you choose the heart, you will go free, and
you may take Medea with you. If you choose the dagger, you
die. It is up to Fate!"

Now, Haroun knew how sneaky Kazir really was. He knew

that Kazir would take no chances—and especially that he would never risk losing his voluptuous Medea! What Kazir had clearly done (reasoned Haroun) was to draw a dagger on *both* slips. So what could Haroun do? He must choose one, and clearly it would contain a dagger. But he could not demand that the other one be opened to verify that it contained a heart; this would be tantamount to doubting Kazir's word, and doubting Kazir's word was another crime punishable by death.

So what could he do? Suddenly he had a clever idea! He took one of the two slips, tore it into pieces, and swallowed them. He said, "This is the piece I have chosen. If you open the other, you will see what it was."

As I say, Haroun was clever. But not clever enough! Kazir had anticipated that Haroun would do precisely this, hence he had drawn a *heart* on both slips of paper! So when the other one was opened, and seen to contain a heart, this proved to the court's satisfaction that Haroun had chosen the dagger.

So Haroun was now a condemned man. However, Kazir said: "In my country, a condemned man is always given a one-third chance of escape. This privilege will be extended to you. You must choose among three Black horses.

"One of the three Black horses—the newest one—is an enchanted horse; the other two are normal. You will mount the horse of your choice, and it will take you into the desert. If you have chosen one of the two normal horses, you will never make it out of the desert; both you and the beast will perish. But if you have chosen the enchanted horse, then at some point in your journey, it will suddenly sprout wings, fly up in the air, and carry you home to safety. So choose, Haroun, and you had better choose well!"

So spake Kazir. Haroun carefully considered the matter. Obviously the "horses" in the position were knights, and "newest" probably meant "promoted"! So the problem was to decide which of the three Black knights was a promoted one. This Haroun could not remember directly. But he did remember two vital things—namely, that no *White* piece now on the board was promoted, and that the pawn on b3 came from b2. From this, Haroun deduced which was the promoted horse.

He chose accordingly and mounted the horse, which, sure enough, took him into the desert. Hours went by, and the horse appeared to be going slower than ever. At one point, Haroun began to worry that he might have made a mistake in his reasoning. But just then, the horse sprouted wings, soared with him up into the air, and carried him home to safety.

Which horse did Haroun choose?

⁊⁊ ⁊⁊ ⁊⁊ ⁊⁊

Black–15

White–14

According to another version of this story, the position was the one above. And the given side conditions were that all White pieces on the board are original, and that the White king has never moved.

Which horse should Haroun choose this time?

The
STORY
of
AMELIA

✧ 45 ✧

CASE OF THE
ACCUSED QUEEN

Haroun had just come back from the wars (or else it was a long vacation, I forget which) and had not seen his beloved queen for a long time. I won't go into how much he anticipated his homecoming, and especially his first night in the harem; I will only say that during the night in question, he sensed (or was it his imagination?) that a strange and subtle change had come over his beloved! Gradually the thought dawned upon him, nay haunted him, that perhaps she was really a different person! She hastened to assure him that this was nonsense, and that it was *he* who had changed and consequently fancied that she was different. However, the more he brooded over the matter, the more he could not help but believe that his intuition was correct, and that she was a promoted queen, an impostor who had usurped the real queen's place. Finally, in desperation, he brought this problem to his oracle in Mecca (who, incidentally, was the grand vizier in disguise and, in this instance, sounded remarkably like a modern-day psychoanalyst).

"So," said the oracle, "you fancy that Queen Amelia has changed?"

"Two objections," said Haroun. "The first is that your use of the word 'fancy' totally begs the question."

"I have not been brought up in logic and word games," interrupted the oracle. "I deal only with cold, hard psychological facts."

"My second objection," said Haroun, ignoring this interruption, "is that I don't think Amelia has changed. I believe she has been captured—and heaven knows what has happened to her!—and that it is another person—a promoted queen—who

99

sits on her throne. That is very different from saying that Amelia herself has changed."

"More word games!" said the oracle. "You have a real psychological problem, and you insist on wasting time with all these semantic, logical, linguistic, philosophical irrelevancies! Now, has it ever occurred to you that your fear that Amelia has changed may be a subconscious *wish?* Perhaps deep down you are tired of Amelia, and really *would* like to have another queen!"

"Nonsense!" shouted the caliph.

"Ah!" said the oracle triumphantly. "Your very anger only proves my point!"

"It doesn't!" said Haroun.

"It does," said the oracle.

"It doesn't," said Haroun.

"It does," said the oracle. "Anyway, do you have *objective evidence* that she has changed, or is it only a matter of your own subjective feelings?"

"No objective evidence," admitted the caliph sadly, "and that is the whole trouble! If I had objective evidence, I would simply hang the false queen and be done with it! I wouldn't have to come to *you!*"

"Then I don't understand. Do you believe the present queen is an impostor or not?"

"I'm *sure* of it!" cried Haroun.

"Then why don't you hang her?" asked the oracle.

"Because I'm afraid that I may be *wrong!*" groaned the caliph.

"Interesting case," said the oracle, as he wrote down in a large brown notebook: "Supplicant is *sure* that he's right, but *afraid* that he's wrong."

"Don't you see?" cried Haroun. "I was brought up in an ultrarationalistic tradition. My tutor was a Greek philosopher who was always warning me that intuition, no matter how strong, is totally unreliable unless checked by careful rational verification. As he used to say: 'Intuition without reason is as capricious as the south wind.' "

"What's so capricious about the south wind?" inquired the oracle.

"Oh, I don't know," said Haroun wearily.

"Very interesting case," repeated the oracle as he wrote again in his large brown notebook: "Supplicant displays typical schizophrenic conflict between intuition and reason."

"I wonder what I should do," said

Black–13

White–11

the caliph. "Maybe we should consult the chronicler?"

"Impossible," replied the oracle. "The chronicler no longer functions."

"What?" cried the caliph, jumping up from the divan.

"No," said the oracle, "he has for some time now withdrawn completely from all worldly affairs. These days he just sits in meditation—dead as a rock. I don't think any force in the universe could move him."

"Good grief," said Haroun. "This is terrible! What can I do?"

"Why," replied the oracle, "it is obvious what you must do; *you* must be the chronicler!"

"I can't," cried Haroun. "I have never chronicled a day in my life!"

"Ah," replied the oracle with a smile, "but now you have to! Tell me, Haroun, what do you know about the history of the game?"

"Not one bloody fact!" said Haroun.

"That is not true," responded the oracle. "You know *all* the facts, but unfortunately only on an unconscious level. Tell me, have you not been informed of all the moves that have been made?"

"Of course, I am informed of every move that is ever made. I have been *informed* of all the moves of this game, but I don't *remember* any of them."

"Which exactly proves my point," said the oracle triumphantly. "You have repressed all those memories! Now Haroun, that is where *I* come in. My task is to aid you in unearthing all your buried memories—some of them perhaps going back to your early childhood—of the events of this game. Our hope is that you will unearth sufficient data to give an *objective* basis for an evaluation of who really sits on the queen's throne."

"How do I do that?" asked Haroun, none too eagerly.

"Well," said the oracle, gleefully rubbing his hands, "we have a marvelous new technique known as *free association*. You simply lie down on the divan and say whatever comes to mind. One unconscious memory after another will come to the surface until you reach the ones necessary for the solution of your problem. The process may take weeks, months, years, decades, or possibly several lifetimes, but if you truly love Amelia, as you claim, you surely will not flinch from the task."

And so started Haroun's analysis. It did not take him several lifetimes, nor decades, nor years, but only a few months. The relevant facts that came to the surface were: (1) Neither the

White king's nor White queen's knight ventured beyond the sixth row. (2) The Black royalty have never moved nor been under attack. (3) The Black king's knight has moved only once.

From these facts it is possible *objectively* to deduce whether Haroun's suspicions are justified.

Are they?

46

WHICH QUEEN?

When Haroun came back* from his "analysis," joyful at having resolved his problem and gleeful at the prospect of hanging the false queen (or perhaps of devising a choicer form

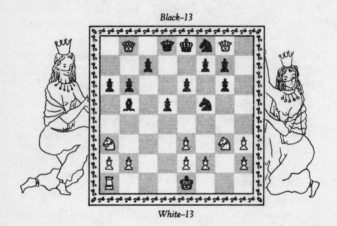

Black–13

White–13

of amusement), almost everything had changed. After all, it took time for Haroun to recall consciously the facts he had subconsciously known about the placement of the pieces, as well as the pertinent points in the history of the game. And not only did he have to unearth the facts, but then there was the task of correlating them and making the relevant deductions. This took longer than expected. So, as I have indicated, much

---

* Strictly speaking, Haroun's absence may appear incompatible with the conditions of chess. But it is to be remembered that when Haroun is away on a nonmilitary venture, a surrogate king is appointed to take his place.

water had flowed under the bridge! For one thing, the real Amelia had finally evaded her captors (for she had been only kidnapped and not killed) and had returned. Then an entirely new battle had begun. What was Haroun's surprise to see *two* White queens on the board!

One of the White queens is Amelia, and the other is precisely the same impostor as in the preceding story. Needless to say, she wishes to retain the false queenship as doggedly as ever. Besides, Haroun happens to be her "type," and she does not want to give him up. She recalls only too vividly those few precious nights of orgiastic splendor. And, of course, the impostor queen is so adept at the art of impersonation that she is virtually indistinguishable from the real Amelia.

Ah, poor Amelia! the loving, toiling, self-sacrificing wife who had cooked and slaved for years to please her Lord and Master, only to be kidnapped one day and removed from her native soil to serve as a slave to other men's lusts. All this she endured with the stoicism typical of her lofty nature. But to come back and find a brazen young hussy claiming to be the *real* queen, to be robbed of her birthright, to be robbed of the man she had been willing to slave for, this was too much! Her noble spirit was at the breaking point!

So, Haroun is again in a fix. Which lady should he take to the harem, and which to the gallows? All those months of cerebration wasted! Thereupon Haroun returned to his oracle in Mecca to help him remember the facts to solve this latest problem.

The facts he recalled were:

(1) Black's first move was the pawn from d7 to d5.

(2) The Black royalty has not yet moved, and White can castle.

(3) The square d7 has been occupied only once.

What is the solution?

47

A NEW
COMPLICATION

Haroun, happy once again, returned to the scene of action, armed with the crucial knowledge that would guide his future life. But hang it, the idiot (I refer now

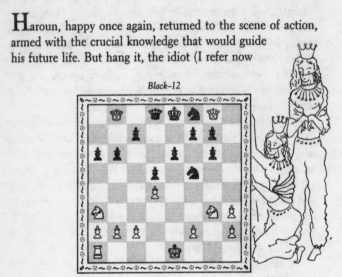

Black–12

White–13

to Haroun) seemed to forget that when one withdraws from the world of action and remains suspended in the world of thought, things happen in the meantime. And so in this instance an entirely new battle—similar in many respects, but not quite identical with the old—had begun in his absence. Again both White queens are on the field, possibly in the same places as before, and then again, possibly interchanged.

So Haroun was once more in a quandary and had to go back to Mecca to consult his oracle. But this time, he determined to manage things a bit better: He left the strictest injunctions that in his absence, no White subject—and especially not the queens—should make any moves.

His pilgrimage to Mecca to seek the advice of the oracle brought out these relevant facts about the history of the game:

(1) The game commenced with a queen's pawn opening: White's first move was with the pawn on d4 from d2; Black countered with a pawn from d7 to d5.

(2) The Black royalty have not yet moved and White can castle.

(3) The squares d7 and e2 have never been reoccupied nor traversed.

Which of the White queens is Amelia?

48

THE RESCUE
OF AMELIA

Upon returning to his kingdom, Haroun found things again
in a totally changed condition.

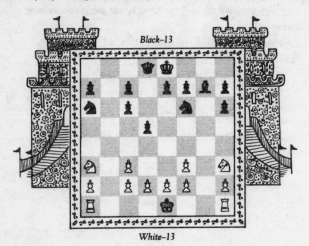

Black–13

White–13

Indeed, there are now *no* White queens on the field! Haroun
had forgotten that his strict orders not to move had no power
over the Black forces, which had immediately moved in and
annihilated the immobile Whites. Hence one battle had been
lost by Haroun and another had started in his absence.

One of Haroun's most loyal subjects informed him of what
had happened in his absence. It appears that nothing was
known about the fate of the impostor queen. In the first battle,
she was either slain or taken prisoner, or else just slipped away.
In the battle then in progress, however, Amelia was taken
alive by a Black castle during a sudden and unexpected on-

slaught of the Black forces. Whether she was captured by the king's or queen's castle was not known or remembered. But Amelia was imprisoned in the castle, and is still there. Both Black castles were themselves subsequently captured by the Whites (as the reader can see, since neither is on the field).

Now, when a castle is captured, it is either buried on the spot, or else it is transported from the field to one of the four borders (just as when chess is played on a chess table). How are castles transported? Good question! There are two methods of transporting a castle: (1) by labor; (2) by magic. The first method is less costly but more arduous; the second is more pleasant but more expensive. In this game, as it happened, the two Black castles were transported by magic. The king's castle was sent to the east border and the queen's castle to the west border. Amelia, of course, is imprisoned in one of them. Haroun plans to start out at his earliest convenience to make a journey to each of them in order to find out in which one his beloved Amelia is imprisoned and to bargain for her release. How ironic, thought Haroun, that I was so impatient with the vizier's romantic *Tale of the Lady and the Knight* and thought it overly sentimental, yet here I am in the very same position myself! Ah well, I really must start out soon to rescue Amelia; I can't let too much time go by.

Just at that moment, a page came rushing madly into court, knocking over all who were in his way. He rushed up to the caliph, gasped "Haroun," and fainted dead away at his feet. The whole court fell all over themselves in their efforts to revive him. When they finally did, he jumped up and continued: "Haroun, I bear excruciatingly important news, there is not a moment to be lost!" And he became so excited that he again fainted dead away. When revived again, he continued: "Haroun, the Black king has issued an edict of the utmost importance to your majesty concerning Mrs. Haroun Al Rashid. It seems that the Black king feels that your Amelia has been demoralizing the entire castle with her moralistic lectures, hence he has decreed that she be executed, and that no later than six o'clock tomorrow morn." At this the page once more fainted dead away, and this time no efforts to revive him were successful.

Haroun jumped to his feet with the knowledge that he must start out almost immediately to save his loved one. He couldn't possibly make *both* castles before tomorrow dawn. Hence he must decide, and within the hour, to go east or to go west. Since he doesn't have time to go to Mecca for advice, he must make his own decision! Fortunately for the history of the world (which would have been quite different if events had turned out differently than they had), Haroun had reached a psychological state where he was no longer dependent on his oracle. Thus he calmly appraised the situation, gathered all relevant historical data, and within half an hour deduced the whereabouts of his virtuous Amelia. He then journeyed forth in the correct direction.

The only historical data relevant to the solution is that none of the remaining royalty has moved. And, as you already know, the White queen was captured by a Black castle. The problem, then, is which one?

◈ 49 ◈

ADVENTURE IN THE FOREST

When Haroun reached the Black king's castle, the Black king fell upon his old enemy and embraced him with tears of joy. "Haroun, old boy, you have no idea how glad I am to see you," he blubbered. "You have come just in time! I was very worried that if I really had your Amelia executed as planned, she would be immortalized in the minds of my men as a martyr, and then her ideas would really take hold of their superegos and I would soon have no army left! Take your dear sweet thing away, and I will give you twelve of my choicest females into the bargain!" Haroun thanked his enemy warmly, and assured him he would take Amelia away, but declined his offer of the twelve females, explaining that he had no mind for that sort of thing at the moment.

And so Haroun was royally wined and dined at the Black king's palace. After the magnificent banquet, a royal entertainment began, which culminated in Kazir's most celebrated chess masters improvising some royal chess problems dedicated to Haroun and commemorating the temporary friendship of the two kings. (These same problems can be found in Appendix I, on page 119 of this book.)

Upon the morrow of this memorable night, Haroun and Amelia departed, after Kazir promised to pay Haroun a return visit. All day long Haroun and Amelia walked slowly homeward. The journey seemed to take ever so much longer than it had in coming, and at twilight they found themselves in an unknown forest, still quite far from home. So they decided to put up at a White castle. Now, the trouble was that they couldn't put up at a *queen's* castle, since men there are executed on sight, nor could they put up at a king's castle, since king's castles are inhabited entirely by men, and any woman who dared to intrude would be instantly attacked.

Fortunately for the royal pair, however, a *third* castle was in sight. Just built, it belonged to neither the king nor the queen and was not governed along the archaic lines of the other two. In short, it was a *co-ed* castle, and there the royal pair would be most warmly received and

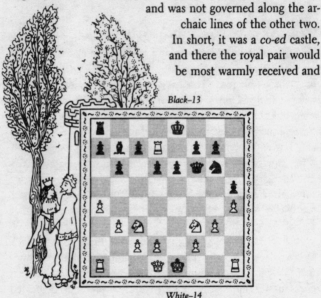

Black–13

White–14

entertained. The problem was to decide which of the three White castles was the Elysium.

So, given that neither king has moved, which of the three White castles is the promoted one?

IN SEARCH OF
THE WATER OF LIFE

After a heavenly orgy in the promoted castle, the royal pair departed in the morning. They arrived home by evening, amid the joyous shouts of all their subjects. Chess problems were composed right and left.

Toward midnight, Amelia began to complain of a strange ache in the small of her back. As the hours advanced, this pain began to spread to other areas, and a fever set in. By morning her entire body was affected, her fever had mounted to 112 degrees, and she was in a coma. Haroun frantically summoned the wisest physicians of the kingdom to her rescue, but none of them could diagnose her condition until a doctor from an eastern province recognized it as jubjubitus—a rare condition caused by the bite of the jubjub bug. These jubjub bugs were known to exist in the cellars of the Black king's castle where Amelia had been confined. "And what is the cure?" asked Haroun eagerly. "There is no cure known to science," was the cheerless reply. "However," continued the doctor, "that does not mean that there is no cure known to *magic*. Now it so happens that I hold a degree in magic as well as in medicine, hence I may be able to help you.

"You see," continued the doctor, "the jubjub bug has its entire being grounded in the evil spirit of Jubjub. However, Jubjub has an archenemy, Frubfrub—likewise a spirit—who is constantly nullifying the work of Jubjub. What we must do is to conjure up the spirit of Frubfrub."

So saying, the physician took out a small vial of green fluid and spilled the contents upon the marble floor. Immediately a thickish green vapor arose, in the midst of which a weird green face appeared, frowning at Haroun. "And why dost thou disturb me from my slumber, O mighty monarch?" asked the spirit. Haroun knelt on one knee and said: "Exalted Frubfrub, thy enemy Jubjub has a victim in his power. She will soon die unless I have benefit of thy wise counsel." Frubfrub knotted his brows and thought for a moment. Finally he said: "Time is short, the vapor thins. Listen carefully, O Ruler of the Faithful. There are four castles now on the field. In each of their centers lies a small black circular room. In the center of each room lies a small white vial, and each is guarded by seven deadly spiders.

Black—15

White—14

Of the four vials, two contain a few precious drops of the water of life. But each of the other two contains a few drops of the water of death. You must locate the two vials containing the water of life; together they are enough to save Amelia. But if either one—or both—should contain the water of death, the case is utterly hopeless."

"And which two castles contain the water of life?" asked Haroun desperately.

"The ones that have not yet moved," was the apparently none-too-helpful reply.

"Yes, yes," cried Haroun, "but which are the two that haven't moved?"

The vapor was thinning and the spirit dissolving. All Haroun heard, very faintly, was: "The Black king has never moved, and the White king has moved only once."

So, the Black king has never moved, and the White king has moved only once. Two of the castles have never moved, and they contain the water of life. Which two are they?

This seems like a rather good place to end our story. Haroun (aided by the vizier) was able to locate the water of life and bring it back to Amelia, who recovered instantly. The two lived quite happily ever after.

Haroun finally signed a disarmoring treaty with Kazir, and henceforth no knights played any pranks for fear of being disarmored.

As for the vizier, after teasing Haroun with a few more thefts from the treasury (which he did primarily out of a spirit of mischievous fun), he settled down and became a marvelous friend of the family. He was a particular favorite with the grandchildren of Haroun and Amelia, of whom I am the youngest. As a little child I sat for hours on his knee listening to all these enchanting chess tales. Fortunately, I learned to write at an extremely early age, and have recorded all I can remember for the entertainment of my readers.

Peace be to all beings!

APPENDIXES

APPENDIX I

PROBLEMS COMPOSED
AT KAZIR'S CASTLE

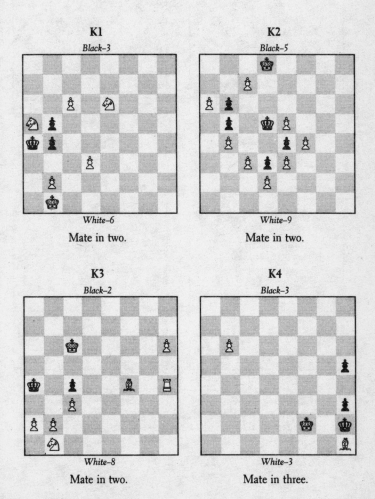

K1
Black–3

White–6

Mate in two.

K2
Black–5

White–9

Mate in two.

K3
Black–2

White–8

Mate in two.

K4
Black–3

White–3

Mate in three.

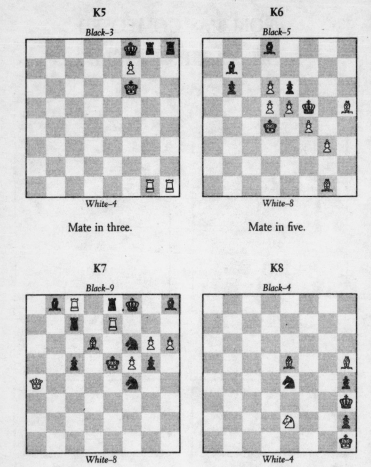

K5
Black–3

White–4

Mate in three.

K6
Black–5

White–8

Mate in five.

K7
Black–9

White–8

1. Self-mate in two.
2. Same with Q and P (g6) removed.

K8
Black–4

White–4

Self-mate in three.

K9

Black–7

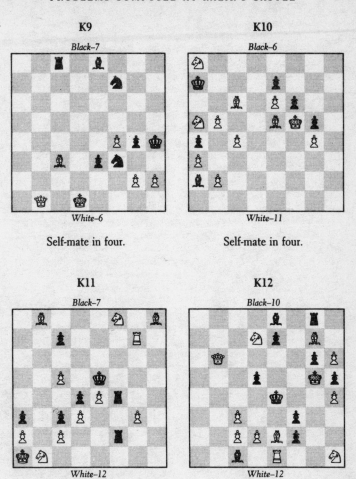

White–6

Self-mate in four.

K10

Black–6

White–11

Self-mate in four.

K11

Black–7

White–12

Self-mate in four.

K12

Black–10

White–12

Self-mate in five.

APPENDIX II

SOLUTIONS

It is tempting to give the following fallacious argument that the position is quite impossible regardless of where the White king stands:

"Either the White king stands on b3 or Black is now in check from the bishop. But the White king can't stand on b3, for if it did, it would be in imaginary (impossible) check from the Black rook and bishop—i.e., it would be simultaneously in check from these two pieces, and whichever just moved to check it, the other one would have been checking it before the move. Therefore the White king is not now on b3, and Black *is* in check. How did White give this check? Not by moving the bishop; hence the White king must have moved from b3 to discover the check (it obviously couldn't have moved from c2). But if it moved from b3, it would still have been in an imaginary check from the Black rook and bishop. Therefore the position is impossible."

It is amazing how many people will swear by the validity of the above argument! But it is wrong, and for the following reasons. It is indeed true that the White king cannot stand on b3 *now*, although it does not follow that it could not have just moved from there; it might have just captured a Black piece. How does that help? Well, put the White king on b3 and a Black pawn on c3—this was the situation before the last move. Now how did Black just check White? Only by

the Black pawn on c3 coming from b4 and capturing a White pawn on c4 *en passant.* So, put the Black pawn back on b4 and a White pawn on c4—this was the position the move before that. Going back another move, put the White pawn back to c2. Going back one more move, put the Black bishop anywhere on the diagonal from a8 to h1—say, e4. Then the position was this:

The following sequence of moves then brings the game to its present position:

	White	Black
1.		B-d5 ch.
2.	P-c4	P × P e.p.
3.	K × P	

Hence the White King now stands on c3.

2 • INVISIBLE BUT NOT INVINCIBLE!

It is White's move, so Black is not in check. Hence the Black king does not stand on d8. This means it is either on c8 or White is in check from the Black rook. If the latter, then the Black king must have just moved from c8 to d7. Therefore the Black king is now either on c8 or d7. If the White pawn on c7 now captures the Black rook and becomes a knight, then the Black king will be mated no matter on which of the two squares he stands.

3 • HAROUN IN DISGUISE

Haroun must be the Black pawn on c7. It is easiest to prove this by *reductio ad absurdum*—i.e., to assume he isn't until this is contradicted. Suppose he is not the pawn on c7. Then he is disguised as something else. He cannot be any of the three pawns on a7, a6, or b5, because he would then be in check at the same time Black is. Therefore these three pawns are for real. The pawn on a6 came from b7, making one capture, and the pawn on b5 came from d7 making two captures. This totals three captures, *all on white squares*. Hence the White queen's bishop, which travels only on black squares, was captured separately. Hence four White pieces have been captured, so there can be at most twelve White pieces on the board. This means (still assuming that Haroun is not the pawn on c7) that Haroun must be disguised as a *White* piece, for if one of the Black pieces were really Haroun, then there would be thirteen White pieces now on the board. Thus we have proved Proposition 1: *If Haroun is not the Black pawn on c7, then he is one of the White pieces.*

Now, continuing the assumption that Haroun is not on c7, which of the White pieces could he be? Well, he cannot be on a4, b7, c6, c4, d7, f7, g8, or g6, since on any of these squares he would be in check (recall Black is in check). Hence he is one of the four pieces on d5, e4, g4, h5. Thus we have proved Proposition 2—*If Haroun is not on c7, then he is on d5, e4, g4, or h5*. Now (still assuming that Haroun is not on c7) comes the interesting part of the analysis!

The White queen has the Black king in check. White's last move must have been with the queen coming from f8 and capturing a Black piece on g8. So let's go back a move; put the White queen on f8 and a penny on g8 to represent the unknown Black piece just captured by the queen. What was Black's last move? It couldn't have been the king from g7, since White would have had to get to f8 to check him on g7 with the queen. Nor could any of the Black pieces on the board, other than the penny on g8, have moved last, since they are all bottled up (the only exception is the pawn on b5, but even it could not have just moved from b6 because it came from d7 via c6). So the Black piece on g8 moved last. It cannot be a bishop, so it must be a queen, rook, or knight. Whichever it is, it just moved from somewhere to g8 *without capturing a piece* (since g8 is a white square, and all three missing White pieces eligible to be captured on white squares have been captured by the pawns on a6 and b5). So the Black piece on g8 just moved there without making a capture. If it is a queen or a rook, then it must have come from g7, which is impossible since the White queen couldn't have gotten in to check the Black king. It must be a knight. This knight couldn't have come from h6, because again the queen would have had no square to come from to check Black. So the knight came from f6 (thus enabling the White queen to come from h6 to check Black). Since on Black's last move, a knight came from f6, then none of the four pieces on d5, e4, g4, and h5 could really be Haroun, since each of these four squares was in check from the knight on f6! Yet, if Haroun is not the Black pawn on c7, he must be one of these four pieces (by Proposition 2). So the only way out of the difficulty is that Haroun must be the Black pawn on c7.

On the second night, Haroun is not a Black pawn (since he is costumed differently). Then, by the same argument as before, Haroun must be White, and must be either one of those same four pieces (d5, e4, g4, h5) or else the pawn on b7 (which he can be, since the Black queen is no longer there to check him). And again by the same argument as before, he cannot be on d5, e4, g4, h5. Therefore he is disguised as the White pawn on b7.

Step 1: The Black bishop on a2 cannot be original, since the White pawn on b3 would have prevented its getting there. Hence this bishop is a promoted bishop. The Black pawn that did the promoting must have come from e7, captured four pieces to get to a3, then have gone to a2, and then made a capture on b1, where it promoted. Thus the pawn from e7 has made five captures. The White bishop originally from c1 never left its home square (since neither of the pawns at b2 or d2 has yet moved), and hence was captured on c1. This makes six captures of White pieces. Therefore the enchanted rock on g4 must be a Black piece.

Step 2: White's last move could not have been with the rook from e1, where it would have checked Black, nor with the king (which could have come only from b1, where it would have been in an impossible check from the bishop), nor could it have been with any piece other than the rook or king. Therefore White has just castled. Hence the White king never moved before that.

Step 3: Among the White pieces captured by the Black pawn from e7 en route to promotion was the White rook from h1. Since White has just castled, and the White king never moved before that, how did the rook from h1 get onto the board to get captured by the pawn? The only possible explanation is that the White pawns on g3 and h3 have cross-captured to let out the rook—i.e., the pawn on g3 really comes from h2 and the pawn on h3 comes from g2. Now, since the pawn on g3 comes from h2, then the Black bishop on h2 has always been confined to the two squares g1 and h2. How did the bishop ever get to either of these two squares? The only explanation is that the Black bishop on h2 is also promoted!

Step 4: The promoted Black bishop on h2 must have been promoted on the square g1. The pawn which promoted on g1 must have come from g7, since neither of the pawns from f6 or h6 could make a capture to get on the g-file (all six missing White pieces have already been accounted for), and the pawn from e7 has promoted to the bishop on a2. What happened was this: The White pawn from g2 made its capture on h3 while the pawn on g3 was still on h2; this left the g-file open for the Black pawn to come down and promote (but after the White rook from h1 got out), and then the pawn on h2 made its capture on g3.

Now we are ready to ascertain the identity of the enchanted rock. We already know it is Black. It can't be a pawn, because the pawns from g7 and e7 have promoted to bishops, and the pawns from f7 and h7 could make no captures to get to the g-file. Also it is not a queen or a rook, since there cannot have been any more Black promotions (the pawn from f7 could never have gotten past the pawn on f2, and the pawn from h7 was always blocked by a White pawn on h2 or h3). Therefore it is a bishop or a knight. However, White has just castled and the White king has just passed over the square d1, so the unknown on g4 cannot be a bishop because a bishop would have checked the square d1 (and in castling, the king may not cross a checked square). Therefore the enchanted rock must be a Black knight.

As for the second part of the problem—whether or not Black can castle—we continue our analysis as follows.

The four missing Black pieces are the two original bishops and the pawns from f7 and h7. Three of these have been captured by the pawns at b3, g3, and h3. The pawn from f7 is the piece which has not been captured by these pawns (since it could not capture to get off its file). Now, White is missing six pieces; five of them have been captured by the pawn from e7 en route to promotion on b1, and the sixth is the White queen's bishop which fell on its own square. Therefore the White pawn from e2 was either one of the five pieces captured by the Black pawn from e7 or else this White pawn has promoted. If the former, then the White pawn must have gotten to the left of the e-file, but this is not possible since the only piece it could have captured is the pawn from f7. (Recall this pawn didn't promote.) Therefore the White pawn from e2 did promote. It either promoted on e8 without making any captures, or it promoted on f8 after capturing the pawn from f7 somewhere on the f-file. If the former, then the Black king has previously moved. If the latter, then the White pawn must have passed through the square f7 and made the Black king move. Therefore the Black king has moved, and Black cannot castle.

5 • THE HIDDEN CASTLE

Step 1: The pawn on f6 has come from e7 and the pawn on e6 has come from d7. What White pieces were captured on e6 and f6? It couldn't have been a White knight on either square, since neither the Black king nor Black queen has ever moved or been under attack. Also, the missing White queen's bishop was captured on its home square c1, so could not have been captured on f6 (nor, of course, on e6). This means that unless the missing White pawn from h2 has promoted, the pieces captured on e6 and f6 must be the missing White castle (only one is missing, since the other is somewhere on the board, although we don't know where) and the pawn from h2. However, the pawn from h2 cannot have been captured on e6 or f6 for the following reasons: In order to get even to the f-file, it would have had to make two captures. The Black queen's castle never got out onto the board to be captured by the White pawn from h2, because the Black queen has never moved, and hence the castle was captured on a8, b8, or c8. Therefore the only way the White pawn from h2 could get to the f-file is by capturing both Black bishops. But the bishops couldn't have gotten out till *after* the pawns on e6 and f6 had both made their captures! Hence the pawn from h2 did not get captured at e6 or f6. Therefore the pawn from h2 has promoted.

Step 2: In order to promote, it had to get past the Black pawn on h6. The only way it could have done this is by first capturing one of the Black bishops somewhere on the g-file and then capturing the other one on h7. After this, it had no more pieces to capture (since the Black queen's rook never got out), hence it made its promotion on h8. Since both Black bishops were captured prior to the promotion, both pawns on e6 and f6 made their captures *prior* to the promotion (to let the bishops out). Hence the promoted White piece was *not* captured at e6 or f6. Hence two *original White officers* were captured at e6 and f6.

Step 3: What officers were they? Well, one of them was an original White castle. And the other? Not the bishop from c1; not the White queen (since none of the royalty has moved). Also not a knight, as we already know. It could not be the bishop from f1 (a white square) because then the bishop on c4 would have to be promoted and White promoted on h8, which is a black square. Therefore the only possibility is that *both* original White rooks were captured on e6 and f6. Yet there is a White castle somewhere on the board. So this castle must be the piece that was promoted on h8. It has always been confined to the squares h7, h8, and g8. So it now stands on h7.

6 • A VITAL DECISION

The solution is that the Persians *should* send reinforcements.

Looking at the diagram, it is easy to see that White must be moving *down* the board, because if it were moving up the board the White king could never have crossed the sixth rank (from the bottom) to get to the eighth, since each of the squares is checked by a Black pawn that has never moved. Therefore White is at the top of the board, and White is now in check. That means the final message is that Black moved last, and the first message translates as "Black—

Yes." So the answer to the question of whether reinforcements are to be sent is yes.

7 • MYSTERY OF THE BURIED CASTLES

The solution contains two nice surprises! The Black queen's rook must have been captured on e3. Since the Black king has not moved, then the pawns on b6 and c6 have cross-captured to let it out. They must have captured a rook and the queen's bishop or else both rooks. However, since neither White rook could get out *prior* to the capture on e3 of the Black rook, then the bishop fell first on b6, then the Black rook got out and was captured on e3, then a White rook got out and was captured on c6.

Now, since the Black rook got out *prior* to the capture on c6, the original Black queen's bishop couldn't have gotten out yet, hence must have been captured on its own square to make way for the rook to get out. Therefore the bishop on a6 must be a promoted one!

The promoting pawn must have come from f7 or g7 and promoted on f1. The capture on e3 must have occurred prior to the promotion (as otherwise neither pawn from f7 or g7 could get to f1 without making *two* captures). Also the pawn on g3 had moved from g2 *before* the promoted bishop got out from f1. Therefore, before the promoted bishop left f1, *both* pawns on e3 and g3 were on their present squares. How then did the knight on f1 subsequently get to the square? The only possibility is that this knight is also promoted!

Therefore, the pawns from both f7 and g7 have promoted. The one from g7 must have captured the other White rook somewhere on the f-file, and since both White castles were captured on the same row, it must have been f6. Thus the White castles are buried on the squares c6 and f6.

8 • CASE OF THE DISPUTED CASTLE

To begin with, one of the Black bishops on a2 and b7 is obviously promoted. The other one must be original, since there are seven Black pawns now on the board.

It will be shown that the castle on a5 must be Black by first proving the following two propositions:

Proposition 1: If the castle on a5 is White, then the bishop on b7 is the original one.

Proposition 2: If the castle on a5 is White, then the bishop on a2 is the original one.

From Propositions 1 and 2, it of course follows that the assumption that the castle on a5 is White leads to a contradiction; hence the castle is Black.

Proof of Proposition 1: Suppose the castle is White. Then White is missing the bishop from c1 and the pawn from e2. What was captured on a6 by the Black pawn? Not the bishop from c1 (which moves on black squares), nor the pawn from e2 (which couldn't have captured enough pieces to get to a6). Hence the pawn from e2 must have promoted. The piece captured on a6 couldn't be *original*, because if it were, one of the two White castles, White knights, White queen, and White king's bishop now on the board would have to be promoted, but we are given that there are no promoted White pieces now on the board. Therefore the White piece captured on a6 was the *promoted* piece. Hence the capture on a6 did not occur till *after* the White pawn promoted. Before the White pawn promoted, the Black pawn on a6 was still on b7 and the original Black queen's bishop was still on its home square c8. This means that the White pawn from e2 could not have captured more than two pieces en route to promotion, because before the promotion, the Black queen's rook from a8 was confined to the squares a8 and b8 (by the pawn then on b7 and the bishop then on c8), so could not have been captured by the White pawn. The only two pieces free to be captured by the pawn were the Black queen and the Black king's bishop from f8. And since the White pawn had to capture at least two pieces en route to promotion, it took the queen and a bishop and promoted on g8. It couldn't have promoted on c8 because the Black queen's bishop was there. This means the Black queen was captured by the White pawn en route to promotion. Now, we recall that before the White pawn promoted, a Black pawn was on b7 and the original Black queen's bishop was on c8, and also that the Black king was on his home square e8 (since Black has just castled). Hence the only way the Black queen could have gotten out to be captured by the White pawn is by the pawn from c7 *first* moving to c6 to let her out. This means that the pawn on c6 was there *before* the pawn from b7 got to a6 (because first the pawn from c7 moved to c6, then the queen got out to be captured by the White pawn from e2, then this pawn promoted, and then the promoted piece was captured on a6). This means that the original bishop from c8 has always been confined to the squares a8, b7, c8. Therefore the bishop on b7 must be the original one.

Proof of Proposition 2: We again assume the castle on a5 is White. The Black pawn which promoted to a bishop is the pawn from g7. It must have promoted on a white square, hence not on g1, so it has made at least one capture. Now, the pawn from e2 has promoted to the piece captured on a6, so the only White piece which could have been captured by the Black pawn from g7 is the bishop from c1. Hence, while the pawn on g3 was still at g2, the Black pawn was at g3 and captured the White bishop on h2 and then promoted on h1. Before this, the pawn on b3 must have moved from b2 to let out the bishop from c1! But ever since the pawn from b2 was on b3, the bishop on a2 has been confined to a2 and b1. Therefore the bishop on a2 was confined to those squares *before* the promoted bishop was made. Therefore the bishop on a2 must be the original one.

We have thus proved Propositions 1 and 2, and now see that if the castle on a5 were White, we would get an absurdity. Therefore the castle must be Black.

Discussion: Now that we know the castle on a5 is Black, the following question might be raised: Can we tell which of the bishops on a2 or b7 is the original one? No, we cannot. For on the one hand, since the castle on a5 is Black, then the missing White castle could have been the piece captured on a6, hence the White pawn from e2 needn't have promoted at all. This kills the argument that the bishop on b7 must be original. On the other hand, the fact that a White rook is missing means that it, rather than the bishop from c1, could have been the piece captured by the Black pawn at h2. This kills the argument that the bishop on a2 must be the original one. So either one could be original.

9 · MYSTERY OF THE BLACK CASTLE

White started out with fifteen pieces, and thirteen are still on the board. Hence two have been captured, and they were captured on b6 and g6 by the pawns now there. No other captures of White pieces have been made. In particular, the missing Black pawn from e7 could not have made any captures to get off its file, and it could not have promoted (since the White king has not moved). So the pawn from e7 was captured somewhere on its own file.

Next we observe that the missing White pawn from e2 could not have made more than one capture, for the bishop from c8 was captured on its home square and the pawn from e7 neither promoted nor got captured by the White pawn (because it would have had to make a capture to get off the e-file to get captured by the White pawn). This leaves the missing Black rook as the *only* piece which could have been captured by the White pawn from e2.

After this preliminary survey, it is necessary to determine which two White pieces got captured on b6 and g6. We are given that White gave Black odds of a knight, hence from White's original set of fifteen pieces, the two missing pieces are one knight and the pawn from e2. Either these are the two pieces that were captured on b6 and g6, or else the White pawn has promoted. Now the pawn could *not* have been captured on b6 or g6 because it would have had to make at least two captures to get to either of these squares, and we have proved that it made at most one capture. Therefore the pawn from e2 has promoted. It could not have promoted on e8 because the Black king has not yet moved. Also it did not make more than one capture en route to promotion. Therefore it must have come down to e7 and then captured a Black rook on f8.

The next thing to show is that the promoted White piece is not now on the board. Well, since the pawn promoted on f8, it couldn't have promoted to a queen or rook, or the Black king would have had to move. Hence neither the queen nor either rook on the board is promoted. The bishop on c1 is obviously not promoted. And the bishop on e2 is not promoted, because it is on a white square and the promoted piece was made on f8, which is a black square. Hence no White piece now on the board is promoted. Therefore the promoted White piece was one of the two pieces captured on b6 or g6. This means that at most *one* of the captures on b6 and g6 was made prior to the promotion. Hence at the time the White pawn on e7 captured a Black castle on f8, either the pawn now on b6 was on a7, or the pawn now on g6 was on h7 (or possibly both). If the former, then the Black queen's castle could not get *out* to be captured on f8. If the latter, then the Black queen's castle could not have crossed the square h7 to get *in* to be captured on f8. Hence it was the king's castle which was captured on f8. So the evil castle on e5 belongs to the Black queen, Medea.

10 · STORY OF THE VEILED QUEEN

Either the pawn on c3 captured the original Black queen or it didn't. *Case* 1: It did. Then obviously the queen on h5 is White (for if it were Black it would be promoted, contrary to hypothesis), and also the original Black queen was captured off her file.

Case 2: It didn't. Then what was captured on c3? Not the original queen; not the bishop from c8 (it moves on white squares), nor the pawn from e7, because it would have to capture more than one piece to get to c3. Therefore the pawn from e7 has promoted. Now, the only way it could promote without capturing more than one piece is by capturing one piece on the d-file and then promoting on d1. What piece did it capture? Either the White king's bishop from f1 or else the White queen (in which case the queen on a5 must be Black).

Now, it couldn't have captured the White king's bishop for the following interesting reason: Suppose it did. Then before that, the pawn on e3 must have moved from e2 to let the bishop out. Thus the pawn on e3 was there *before* the promotion on d1. Also the pawn on c3 must have been there before the promotion, because the Black pawn must have passed through the square d2. So before the promotion, both pawns on c3 and e3 were already there. Hence after the Black promoted piece left d1, the White knight on d1 could never have gotten in to that square! Therefore it was not the White king's bishop which was captured by the pawn from e7. So it was the White queen. Hence the queen on a5 is Black, and the White queen was captured on the d-file, which is the queen's file.

Therefore, if the missing queen has been captured off her file, the situation reduces to Case 1 and the mystery queen is White. If the missing queen has been captured on her file, we have Case 2 and the mystery queen is Black.

11 • STORY OF THE PURLOINED TREASURE

White is now in check from the bishop on d8. Black's last move must have been with his king, from f6. On f6, he was in check from the bishop on a1. How did White give this check? Not by moving the bishop, nor by the pawn on a3 making a capture from b2, since the White bishop could not have gotten to a1. Could it have been with the pawn on e6? Well, this pawn couldn't have moved from e5, because there it would have been checking the king. It could have come from d5 or f5 and made an *en passant* capture of a Black pawn on e5. Suppose it did. Then on the move before that, the Black pawn came from e7; which means the Black bishop on d8 could never have escaped from f8 and hence must be a promoted bishop. Going back a couple of moves, there is a Black pawn on e7, and the pawn on g4 must have come from d7 and made three captures. Then the pawn on d3 couldn't have come from a7 or it would have made three more captures—making six, which is one too many. Therefore this pawn came from c7 and made one capture—this with the captures made by the pawn on g4 makes four. But then the pawn which promoted to the Black bishop must have come from a7 and made two captures to reach the Black square a1 or c1. This again totals six, or one too many. Therefore White's last move was *not* with the pawn on e6.

What could the move have been? The only possibility is that the piece that made White's last move and discovered check from the bishop on a1 was captured by the Black king on its last move. And it must have been a White knight, which on the last move came from e5. Therefore a move ago, there were three White knights on the board. Hence a White pawn must have promoted.

Now, suppose the pawn on g3 came from f2. Then the minimum number of captures made by the promoting White pawn and the remaining pawns is seven, as the reader can readily calculate. (If the promoting pawn came from h2, then it had to make three captures to get around the Black pawns on f7, g7, and h7. The pawn on g3 made one; the pawn on a3 made one, the pawn on a6 made two, and the pawn on e6 made one—this totals eight! But we can get by with seven if the pawn on e6 came from h2. It would account for three captures, g3 for one, a3 and a6 for three, and the promoting pawn could have come from d2 making no captures.) But seven captures is still one too many. Therefore the pawn on g3 really comes from h2 (and the total number of captures now necessary is only five).

Thus the pawn on g3 is not guilty.

12 · PURLOINED TREASURE II

Kazir offered the following proof that Barab (the pawn from b7) did promote and that the promoted piece is now on the board.

Suppose Barab didn't promote. Then he was the piece captured by the pawn on c3. To get to c3, he must have captured a White piece somewhere on the c-file. He couldn't have captured the missing White pawn from h2, hence he captured a White officer. And if he didn't capture an *original* White officer, then an original White officer was captured before Barab was captured on c3.

To prove this, suppose Barab captured a *promoted* White officer. Then the pawn from h2 promoted beforehand. It had no Black pieces to capture, hence it promoted on h8 and the pawns on g6 and h6 cross-captured to let it by. More explicitly, the pawn on g6 came from h7, while the pawn on h6 was still on g7; this left the h-file open for the promoting White pawn. Thus the pawn on g6 was there *before* the White pawn promoted. Hence the pawn on g6 captured an *original* White officer. So, before Barab was captured on c3, either he or the pawn on g6 had captured an original White officer. Which proves that an original White officer was captured before Barab was.

What is the original White officer? Before Barab was captured on c3, the pawn on c3 was on b2, hence neither the White queen's bishop nor White queen's rook could get out to be captured. It could not have been the White king's rook, because White can castle. It could not have been the White king's bishop, because if it were, then the bishop on c4 would have to be promoted, which is not possible since the White pawn promoted on h8, which is a Black square. If it were a knight, then one of the White knights now on the board must be promoted. This is not possible, because a promoted knight could never have escaped from the square h8, since the pawn on g6 was there *before* the promotion. The last possibility is the queen. But this is also impossible, because while the pawn on c3 was still on b2, and the White king had never moved, the queen could not have escaped from its home square unless the queen's bishop was first captured on its home square of c1. But this, together with the three captures made by Barab and the pawns on g6 and h6, would add up to four captures, which is one too many.

The entire argument has shown that if Barab had been captured on c3, then the position would be impossible. Therefore Barab has indeed promoted. We next show that the Black piece to which Barab has promoted cannot have been captured on c3.

Suppose it had been. Then Barab promoted *before* the capture at c3. This means that before Barab promoted, the pawn on c3 was on b2. Hence while the pawn on c3 was still on b2, the pawn from a2 first moved to a3 and then Barab made a capture on a2 from b3. The bishop from c1 has always been confined to the squares c1, b2, a1, and thus must have been captured on one of these three squares. So the bishop on h4 must be promoted. Here, again, the pawn from h2 has promoted and the pawns on g6 and h6 have cross-captured. These two captures, together with Barab's capture made on a2 and the capture of the White queen's bishop on a1, b2, or c1, make four captures, which is one too many.

So what really happened was this: An *original* Black officer was captured on c3, and *then* Barab came down and promoted and replaced it. Hence the promoted Black piece is now on the board.

13 • PURLOINED TREASURE III

It is given that both kings have moved only once and that one of the Black pieces is promoted. It should be relatively easy to find out which one.

The promoting pawn came from a7. Since the only missing Black piece was captured on c3, the White pawn on a4 has never been off the a-file. Therefore the pawn from a7 made at least one capture to get to the b-file. It must have captured a rook, since the two White rooks are the only missing White pieces. Now, the White king has moved only once, so White has castled on the queen's side. Before White castled, the queen's rook was still on a1 and the king's rook couldn't get out to be captured by the pawn from a7. Therefore this pawn did not make its capture till *after* White castled. Thus White first castled, then the Black pawn captured a rook on the b-file. This means the pawn couldn't have promoted on b1 because it would have had to cross b2 and make the king move a second time. So the pawn must have captured the second rook on the a-file and promoted on a1.

A promoted knight could not leave a1 except via b3, where it would check the king and make it move again. A promoted bishop could not leave a1 except by first moving to b2, where again it would check the king. Hence the promoted Black piece now on the board is neither a knight nor a bishop, but must be a queen or rook. Remember that the Black king has also moved only once, so that Black also castled on the queen's side, and the king has not subsequently moved. Hence the original Black rooks have always been confined to the eighth row or to the squares d7, g7, and h7. (Since the two missing White pieces were captured by the promoting Black pawn, the pawns on g6 and h6 couldn't have cross-captured to let the promoted Black rook in to d8 or e8.) Thus neither rook on d8 or e8 is promoted. So the promoted Black piece is the queen.

<h2>14 • PURLOINED TREASURE IV</h2>

Black is missing a queen and king's bishop. Since the Black queen was captured on the eighth row, she was not captured on a3. Hence it was the bishop that was captured on a3.

Gary, the promoting White pawn, has come from h2. Since Black can castle, Gary did not promote on h8. Hence he made at least one capture. But he couldn't have made more than one, because Black is missing only two pieces and one of them was captured on a3. Therefore Gary made exactly one capture and promoted on g8. Since the missing Black bishop was captured on a3, it was the queen that was captured by Gary. Hence the queen was captured on the g-file. Since it is given that she was captured on the eighth row, she was captured on g8. Hence the pawns on g6 and h6 have cross-captured to let Gary by. Moreover, the pawn on g6 was there while the pawn now on h6 was still on g7, so Gary came down to h7 and captured the queen on g8.

If the White pawn promoted to a bishop, it could never have left the squares g8 and h7 to get to c4. So the promoted piece is not a bishop. Also it can't be a rook, because the Black rook on h8 has never moved, so a White rook could never have gotten from g8 to h1. Therefore the promoted White piece is a knight or a queen. If a knight, then it couldn't have escaped from g8 via f6 without checking the Black king and making it move, hence it must have escaped via h6. Now, if it can be proved that the pawn on h6 made its capture there *before* Gary promoted, we will know that the promoted piece is not a knight. We prove this as follows.

To begin with, the three missing White pieces have been captured at a6, g6, and h6 by the pawns now there. Since the White queen's rook was captured on the a-file, then it was the piece captured on a6—this is important! Now, before the Black queen was captured on g8, she must have left her home square d8 and gotten there. Since the Black king has never moved, the queen must have gotten out via b7. Before this, the pawn from b7 captured the White queen's rook on a6. Before this, the pawn from b2 captured the Black king's bishop on a3 to let the rook out (the rook couldn't have gotten out except via b2 since the White king has never moved). And before this, the pawn from g7 made a capture on h6 to let out the bishop from f8!

(In other words, the major sequence of moves was this: First the pawn from h7 made a capture on g6 while the pawn on h6 was still on g7. Then Gary came down from h2 to h7. There he patiently rested while the following things happened: First the pawn from g7 captured on h6. Then the bishop from f8 got out and got captured at a3. Then the White queen's rook got out and got captured at a6. Then the Black queen got out and went all the way around to g8. *Then* Gary captured the queen and promoted.)

This proves that both pawns on g6 and h6 were there before Gary promoted. Hence he could not have promoted to a knight, since a knight could not have escaped via h6. Therefore Gary promoted to a queen too.

15 • TALE OF THE WILY BISHOP

The bishop was promoted on g8 and the promoting pawn came from g2 (had it come from e2, it would have made at least two captures and the pawn on e4 would have made two more to get from g2, and this is one too many). Therefore the pawn on f6 made its capture *before* the White pawn promoted. For, suppose it hadn't: Then before the promotion the Black pawn was on g7, hence the White pawn would have had to make two captures to get around it. One of them could have been the Black king's rook, but the other two missing Black pieces are the bishop from f8, which couldn't have left its home square while the Black pawn was still on g7, and the Black queen's rook, which has to have been captured on a8, b8, or a7. Therefore the pawn on f6 made its capture before the promotion.

The Black pawn on f6, then, captured the White queen's bishop (the only missing White piece). Before that, the White pawn on b3 moved from b2 to let the bishop escape from c1. Therefore the pawn on b3 was there *before* the promoted bishop was made. Hence the promoted bishop could never have gotten to a2. Therefore the bishop on a2 is the original one and the promoted one is on g2.

16 · SECOND TALE OF THE WILY BISHOP

Step 1: It is necessary first to prove that the capture on d6 occurred before the capture on a3.

(a) If the Black queen's rook was captured on a3, then the capture on d6 must have occurred first, to let the rook out.

(b) Suppose the capture on a3 was not of the Black queen's rook. Then what was it? Not the Black king's rook, because this got captured on its own square. Also not a Black pawn, because neither of the pawns from g7 and h7 could make enough captures to get to the a-file. Therefore a Black pawn must have promoted. Also, it must be the promoted Black piece that was captured on a3, because if an original knight, bishop, or queen had been captured on a3, then one of Black's knights or bishops or the queen now on the board must be promoted, contrary to the given fact that no promoted Black piece is on the board. Therefore the promoted Black piece was captured on a3.

Now, the White pawn on g3 comes from g2, because if it came from h2, the bishop on h2 could never have gotten there. Hence the promoting White pawn has come from h2. Also the promoting Black pawn comes from h7, because if it came from g7, it would have to have made at least one capture to get around the pawn on g3, and the pawn on g5 has made a capture to get from h7. Since the pawn on d6 has also made a capture, this would make three captures, which is one too many. So the promoting Black pawn comes from h7. Thus, both promoting pawns come from the h-file, and one of them must have captured to get past the other. The Black promoting pawn did not make more than one capture (since one of the two missing White pieces was captured on d6), hence if it captured to get by the White pawn, it must have made its capture on g2 while the White pawn was still on h2 and the pawn on g3 was already on g3. But then the White bishop could never have gotten into h2! So it was the White promoting pawn that captured to get past the Black one; it captured a piece on g6 or g7 while the Black pawn was still on h6 or h7, then it made a second capture of the rook on h8. What piece did the White pawn capture on g6 or g7? It must have been the Black queen's rook, since the other two missing Black pieces are the promoted pawn captured on a3 and the king's rook captured on h8. So the Black queen's rook was captured while the Black promoting pawn was still on h6 or h7, hence before the Black pawn promoted. But before the Black queen's rook was captured, the pawn on d6 made its capture to let out the rook. So the sequence would have been this: first the capture was made on d6, then the Black queen's rook got out to be captured on the g-file, then the Black pawn came down and promoted, and then the promoted Black piece got captured on a3. Hence the capture on d6 occurred before the capture on a3.

Step 2: We have proved that regardless of whether the piece captured on a3 was the Black queen's rook or a promoted Black piece, the capture on d6 occurred before the capture on a3. Now the piece captured on d6 was the White queen. It got captured there while the pawn on a3 was still on b2 and the original White queen's bishop, therefore, still on c1. Since the White king has never moved, the queen must have gotten out via c2, hence the pawn on c3 must have first moved from c2. So the pawn on c3 was there before the queen got captured on d6, which in turn was before the pawn from b2 made its capture on a3. So the pawn at c3 was there before the pawn at a3. This implies that the original bishop from c1 never left the squares c1, b2, a1. Hence the bishop on b2 is the original one.

17 • ARCHIE'S COUP

White is in check from the bishop on c8. Black's last move was not with the king from f5, for it would have been in an impossible check from both White bishops. Black's only possible last move was with the pawn on g3 coming from f4 or h4 and capturing a White pawn on g4 *en passant*. Before this the White pawn had just moved from g2. Therefore *both* White bishops on the board are promoted; the original king's bishop was captured on f1!

18 • A PAIR OF ODD-BALLS

White's last move was with the queen from d8, capturing a Black piece on e8 and checkmating the Black king. What was Black's move before that? Not with the knight from b6 (where it would have checked the king). Also not with a pawn, because the pawn on a6 has come from b7 and the pawn on b5 has come from d7 via c6. Now if either of those pawns had moved last, the White king could never have crossed the sixth row to get to the eighth (c6 is the only square on the sixth row he could have crossed without being in an impossible check from a pawn). No other Black piece now on the board could have moved last. Hence Black's last move was with the piece just captured on e8. Therefore this piece was not a bishop. If it was a queen or rook, then it must have come from e7 and it must have captured a White piece on e8, otherwise the Black king would already have been in check from the White queen on d8. If it was a knight that just moved to e8, it could not have come from d6 (where it would have checked the White king), hence it must have come from f6, and therefore it must have made a capture on e8, since otherwise the Black king again would have been in an impossible check from the queen. Therefore a White piece was captured on e8. Now, e8 is a White square, and the pawns on a6 and b5 have made their collective three captures all on White squares. Thus all four of White's missing pieces (he started with fifteen, since he gave one piece as odds) have been captured on White squares. Hence, if the White queen's bishop had not been the piece that was given as odds, it would have been one of the four pieces that was captured on a White square, but this is impossible. Therefore it was the White queen's bishop that was given as odds.

For the second problem, we note that, as in the first problem, four White pieces must have been captured (one on e8, and three by the pawns on a6 and b5). The pawn from h2 could certainly not have been captured by either of the two Black pawns, nor could it get to e8 without promoting, which we are given it did not. Hence this pawn must have been given as odds.

Note: In the first problem, the White pawn from h2 *did* promote, since it could never have gotten to a6, b5, c6, or a8 to be captured.

19 · CASE OF THE LAZY KNIGHT

To begin with, the square d7 was once occupied by the pawn on d6. Hence it has never been occupied by any other piece. So in particular, it was never occupied by the Black queen's knight. Also the Black queen's knight has moved only twice. Hence the knight on f6 cannot be the queen's knight (or it would have had to come from b8 via d7). So the knight on c5 is the queen's knight, and it got there via a6, not via d7. Also, the square d7 has never been occupied by the Black queen's bishop (since it was occupied by a pawn), but it must have been traversed by the bishop to get to f5. Therefore d7 has never been traversed by any other piece—in particular not by the Black queen. So the Black queen has never occupied nor traversed the square d7. Yet the queen must have been captured by the pawn on b3, since it is the only missing Black piece. How did the Black queen get to b3? Not via d7, hence it must have first occupied b8, then a7, and then occupied or crossed a6. Now, before the queen occupied b7, the knight must have moved to a6, and the knight must then have left a6 before the queen got to b3 to be captured. Therefore the knight on c5 was there *before* the queen got captured and has not moved since. Also the pawn from a7 made its capture on b6 before the queen got out to be captured. Also the queen could not get out till after the queen's bishop left c8, which was not till after the pawn from d7 moved to d6, so the pawn on d6 was there before the queen was captured. Thus all four pieces on b6, c5, c7, d6—the pieces surrounding the White rook on c6—were on their present squares before the capture on b3 and have never moved since. Now, since the White king has never moved, the White queen's castle couldn't get out to go to c6 until *after* the capture on b3. Therefore the rook on c6 must be the king's rook; the queen's rook is now on h1! So, before the king's rook got out to go to c6, the king's knight must have moved from g1. Had it moved to f3, then the king's rook still could not have gotten to c6 (since the knight moved only once and the king has not moved at all). Therefore it is the knight on h3 which is the king's knight.

20 • LAZY KNIGHT?

The Black piece captured on a3 was not the bishop from c8 (a White square), hence either it was the pawn from b7 or that pawn promoted.

Suppose it promoted. It couldn't make nearly enough captures to get to g2, hence it passed through b2. Therefore the capture on a3 occurred prior to the promotion. This means that an original Black piece was captured on a3. But Black now has on the board two rooks, two knights, a queen, and a bishop on Black squares, hence one of these would have to be promoted (since it was anything but a bishop on White squares that was captured on a3). But we are given that there are no promoted pieces on the board. Hence the pawn from b7 did *not* promote.

So the pawn from b7 got captured on a3. Before this it had to make a capture on the a-file. Before this capture, the pawn on a3 was still on b2. Hence the bishop from c1 couldn't get out to be captured. So it was a White rook that was captured by the pawn from b7, and the rook got out while the pawn on a3 was still on b2. Hence it must have been the king's rook, which got out via g2 after moving to the square g1. So the knight on g1 has indeed moved.

21 • WHICH LAZY KNIGHT?

Black is now in check from the bishop on a2. Was White's last move with the pawn on d6 from d5? If so, Black would have had no preceding move, because the pawn on h5 couldn't have come from g6 (where it would have checked the king), nor from h6, nor could the Black king have come from e8 because it would have been in impossible simultaneous check from the queen and rook. Therefore White's last move was with the pawn on d6 from c5 (and not e5, as will be clear in a moment), making an *en passant* capture of a pawn on d6. Before that, the Black pawn had just moved from d7. And before that, the White pawn had moved from c4 to c5 to give check from the bishop (that's why the White pawn must have been on c5 rather than e5). And before that, it was Black's move, which was with the king from e8, where it would have been in check only from the rook, since the Black pawn then on d7 would have blocked check from the queen.

Hence a Black pawn was on d7 before Black's last move. Therefore the bishop from c8 got captured on its own square, since it hadn't time to move out. This accounts for one capture of Black's six missing pieces. The pawns on c3 and g5 account for two more. And the pawn on d6 accounts for the remaining three, since it came from c4 and before that from e2. This totals six. Hence the Black pawn from a7 was captured either by a White pawn or else promoted. The former is not possible, because the pawn would have to have made at least two captures to get to the c-file, which is one too many, considering the two captures made by the pawn on h5. Thus the pawn from a7 promoted, and did so on b1. Hence the knight on b1 has moved and the lazy knight must be the one on g1.

White has moved last. Even if the knight on a1 is White, it couldn't have just moved. Also the pawn on c3 could not have just come from b2, otherwise the Black king could never have crossed White's third rank to get to the first. Therefore the White king moved last. It clearly could not have just come from d3 or e2. Also it could not have just come from e1, since the only way it could have been in simultaneous check from the queen and rook was if a Black pawn from e2 had just made a capture on f1 and promoted to a queen. But this is not possible, since there are eight Black pawns on the board. Therefore the king just came from d1. He must have just captured a Black piece on d2, otherwise the queen could not have *placed him in check*. The Black piece couldn't have been a rook, or the king would have been in an impossible check from this rook and the queen. It couldn't have been a pawn, since there are eight Black pawns on the board. It couldn't have been a bishop, because the original bishop on black squares never moved from its original square f8, and there are no missing Black pawns for a promoted bishop. Therefore it was a knight (and Black's last move was with the queen from d3, where she wouldn't have been checking the king because the knight was in the way). So the second Black knight was just captured on d2, and since there have been no Black promotions, the knight on a1 must be a White one.

23 • THE KNIGHTS WHO CHANGED ARMOR

To begin with, the pawns on g4, h4, and h5 have collectively made five captures and the one on e6 has made one. This accounts for all six of Black's missing pieces.

The Black king is now in check from the bishop on b1. White's last move could not have been with the pawn on e6 from f5, because this would involve two more captures. Hence one of the two knights on a1 or a3 must be White and have just moved from c2. Then what was Black's last move? It was not with the king from g7, since there he would have been in an impossible check from the bishop on h8 (which could not just have promoted, since this would have involved far too many captures). It also was not with either pawn on a6 or h6, since there a bishop would have been locked in on its home square and this would have involved an extra capture. Also, it could not have been with either knight on b8 or d8, since they could have come only from b7 or c6, where they would have checked the White king. Nor could it have been with either knight on a1 or a3, since whichever one was on c2, the other could have come only from b3 (if it were the one on a1) or c4 (if it were the one on a3), and in both cases it would have checked the White king.

Therefore, the only possibility is that the knight on a3 is really White, and has just come from c2, and that the one on a1 is Black and that Black's last move was with a pawn on b2 capturing a piece on a1 and promoting to a knight.

24 • AN ANCIENT PUZZLE

To begin with, the bishop on h2 is promoted, since the original Black king's bishop had to have been captured on its own square. The promoting Black pawn must have come from h7.

The next point to observe is that the Black queen's castle is the *only* Black piece that could have been captured by a pawn; this is so because the Black king's bishop was captured on f8, the Black queen's bishop was given as odds, no knight was captured by a pawn, and the only missing Black pawn has promoted to a bishop. So the Black queen's rook was the only Black piece captured by a pawn; indeed it was captured by one of the pawns on g3 or g4. No other White pawn has made any captures. In particular, neither pawn from c2 nor d2 has made any captures.

What was captured on b6? Not a pawn, because no White pawn could get there without making a capture. Not the White king's bishop (which was on white squares) and not a knight, since no knight was captured by a pawn. Therefore it was a queen or rook, and since White has a queen and two rooks on the board, a White pawn has promoted. The promoting White pawn made no captures, so it must have come from c2 and promoted on c8. The capture on b6 occurred first to let the White pawn by, hence it was an *original* queen or rook that was captured on b6, so the promoting pawn must have promoted to a queen or rook to make up for it, and is now on the board. No other White pawn has promoted, since it would have had to capture to do so. Therefore the knight on c8, even if it is White, cannot be a promoted White knight. Now, this knight came to c8 *after* the promoted piece left there, hence after the pawn on b6 was already *there*. Therefore, since the knight was not promoted on c8, it must have come in from d6. If it were White, it would have checked the Black king, which has never left its square (since it can castle). Hence the knight on c8 is Black and the one on h1 is White.

For the twin problems, the same argument shows that the piece captured on b6 was a queen or rook. And again a White pawn has promoted, also via c7, so the capture on b6 occurred first, and again one of the White rooks or the White queen on the board is promoted. The White pawn that promoted via c7 did not capture on b8, therefore it must have promoted on c8.

Now suppose the knight on c8 is Black. It must have gotten to c8 after the promoted White rook or queen left that square, and it got in via d6, since a pawn was already at b6. This means that at the time the queen or rook was promoted on c8, the pawn now on d6 was still on d7. Hence the Black queen's bishop never escaped from c8, so the Black bishop on a2 is promoted; and the promoting Black pawn had to come from h7. It would have to have promoted on f1 or h1, but couldn't have promoted on f1 without capturing a piece on the first row. Therefore it promoted on h1. The pawn on g3 would have moved from g2 to let the bishop escape from h1, but then the knight on h1 could never have gotten there!

Thus the assumption that the knight on c8 is Black leads to a contradition. Hence the knight on c8 is White and the one on h1 is Black.

Note that there is no problem now with a White knight being on c8. In this problem (where Black gave no odds) there is no reason why a second promotion could not occur on c8—indeed, this must have happened, since otherwise the White knight at no time could get in from d6. So the White knight on c8 is promoted.

25 • THE INVISIBLE KNIGHT

What was captured by the Black pawn on h6? Not the White king's bishop, since h6 is a black square; nor a White pawn, which would involve far too many captures of Blacks, which has only three missing pieces. So a White pawn promoted. The promoting White pawn must have come from c2 (the pawn from e2 would again involve far too many captures by White pawns). It clearly made at least two captures, because of the Black pawn on c5, which was there after Black's opening move of the game. It could not have made more than two captures because the pawn on a3 has captured one Black piece. So it captured exactly twice and promoted on c8 or a8. But it couldn't have promoted on c8, since it would have to have crossed c7 and attacked the Black queen. Therefore it promoted on a8.

We are given that the castle on a5 has moved only once, so it must be the queen's castle. Now, what two Black pieces were captured by the White pawn from c2 en route to promotion? This pawn, together with the pawn on a3, has captured the queen's bishop, king's bishop, and king's castle. The queen's bishop wasn't captured on a3, so the pawn from c3 captured it as well as either the king's bishop or king's castle. Thus of the king's bishop and king's castle, one was captured at a3 and the other by the pawn en route to promotion. Now, neither the rook nor the bishop could get out *before* the capture on h6. This means that the capture on h6 occurred before the promotion *and* before the capture on a3! Since it occurred before the promotion, then an original piece was captured on h6. Also, since the capture on h6 occurred before the capture on a3, the piece captured on h6 could not have been either the White queen's bishop or the White queen's rook, for before the capture on a3, the White queen's bishop was stuck on c1 and neither it nor the White queen's rook could get out to h6. Also the capture on h6 was not of the White king's rook, because the White king has never moved to let it out. And the capture was not of the White queen, which has never moved. So the piece captured on h6 was an original White knight. There are two White

145

knights now on the board (though one is invisible). Hence one of them must be promoted. So the pawn from c2 promoted to a knight on a8.

We recall that the promoting pawn has captured the Black queen's bishop. Before this, the pawn on b6 had moved from b7 to let it out. So the pawn on b6 was there before the promotion. Hence the promoted White knight could never have escaped from a8 via b6, nor via c7, where it would have checked Black, so it must still be on a8!

26 • WHICH IS THE GUILTY KNIGHT?

We prove that the knight on b3 must have moved more than twice.

Suppose it moved only twice. Then it came from b1, moved once to d2, then to b3, and has not moved since. Now, the Black piece captured by the pawn on f3 cannot be the Black pawn from e7, because this pawn would have to make a capture to get onto the f-file. This capture must have occurred while the pawn on f3 was still on e2, but then neither of the White rooks (which are the only missing White pieces) could have gotten beyond the second row. So the Black pawn from e7 has promoted. Since there are no promoted pieces on the board, it was the promoted piece that was captured on f3. So prior to the promotion, the pawn at f3 was still on e2 and thus the bishop on h3 was still on f1. Hence the promoting pawn must have made a capture on d2 from e3, and, of course, after the pawn on d3 had moved from d2. What did it capture? It couldn't have been the king's rook, because the king's rook was locked in by the king's bishop. Hence it was the queen's rook. Now, the queen's rook couldn't get to d2 while the queen's knight was still on b1, nor, of course, while the knight was on d2. Therefore the knight on b3 was there before the queen's rook got captured, hence before the Black pawn promoted, and therefore before the capture was made on f3. And this capture was made before the king's bishop left its home square. Hence the pieces at b3, d3, f3 were all there before the king's bishop ever left home, and have not moved since. Hence the king's bishop could never get to any square other than f1, e2, d1, c2, b1, a2, hence could never get to h3.

This proves that it is impossible that the knight on b3 has moved only twice. Hence the knight on e4 is the one who has only moved twice. So the knight on b3 is the guilty one.

27 • HAROUN'S VIZIER SOLVES A MURDER MYSTERY

The Black pawn on e6 has captured the missing White castle. Suppose it was the king's castle (as it appears). Then, since the king hasn't moved, the pawns on f3 and g3 have cross-captured to let the castle out. Now, the only Black piece free to be taken before the capture on e6 is the bishop from f8 (because the Black queen and her bishop and rook were all locked in). But to free the White king's rook, the capture on f3 must have occurred before the one on g3. This means that the bishop from f8 was captured on f3, a white square, which is impossible. Hence it was the White *queen's* castle that was captured on e6 and the castle on a1 is really from h1!

The sequence was this: First the Black bishop from f8 was captured on c3 while the pawn on d3 was still on c2. The White queen's castle then got out and was captured on e6. This freed the remaining three missing Black pieces. Now, for the king's rook to get to a1, the pawns on f3 and g3 must have also cross-captured, and the Black queen was captured by a pawn. So the knight on d8 is innocent.

28 • MYSTERY OF THE SPY

The only way the position is possible is that the White pieces are really moving *down* the page and that the White knight is really a White pawn that just came from a4, capturing a Black pawn on b4 *en passant*.

29 • SECOND MYSTERY OF THE SPY

Black is in check from the bishop. White's last move was with the pawn on b6, but this pawn did not just come from b5, because it came ultimately from f2. So it must have come from c5 and captured a Black pawn on b5 *en passant*. Before that, the Black pawn had moved from b7. And White's move before that to give the preceding check? The only possibility is that the Black knight on a6 is really a White rook which moved from c6.

As to the second question, since a Black pawn was just on b7, the only way the White king could have crossed the sixth row to get to its present square is if the Black king has moved. Thus Black cannot castle in the future.

30 • STORY OF THE MASTER SPY

The bishop on h4 cannot be the spy, because if it were White it could never have escaped from c1. Therefore it is really Black and the White king is really in check. Hence Black is not in check. This means that neither of the Black knights is really White, nor the Black queen, nor the rook on h8. Also the rook on a8 can't be the spy, for if it were, the bishop on f1 would be genuine and neither White rook could get off the first row to get to a8. Therefore the spy is either the queen on c4 or the bishop on f1. If the former, then Black has two queens on the board and one of them is promoted. If the latter, then the bishop on f1 is really Black and obviously must be promoted since the Black bishop from c8 could never have gotten out. So in either case, Black has made a promotion.

If the bishop on f1 is genuine, we get the following contradiction: The promoting Black pawn must have come to f2 and made a capture on e1 or g1. If the promoting pawn came from c7 it made three captures to get to f2 and then a fourth. If it came from e7, then it made two captures and the pawn on e5 another two to get from c7, so four captures are still involved. Therefore four White pieces have been captured by pawns. But this is not possible, because White is missing two rooks, a knight and a queen (the queen on c4 is Black by assumption, if the bishop on f1 is White), and only one of the rooks could have been captured by a pawn (on e1 or g1); the other could not get out. Thus the bishop on f1 is not genuine. So it is the spy.

Remark: There is no difficulty now with a Black pawn promoting to a bishop on f1, for the White bishop from f1 was captured on its own square and both White rooks were free to get out. Also, since the promotion was on f1 rather than e1 or g1, only three captures need be involved rather than four. So the three captures were of both White rooks and the missing knight.

31 · TRIAL OF THE BISHOP

Black could not have just moved either knight from a square not checking the White king. Hence Black's last move was with the pawn on f5. Suppose he had moved it from either e6 or g6. Then it previously made another capture to get from f7. Also the pawn on h6 has made a capture. This makes three captures. Then the White queen and both White rooks were these three captured pieces. Hence, since the White king has never moved, the White pawns on b3 and c3 must have cross-captured to let out the queen's castle. These two captures plus the one by the pawn on h3 is one too many, since neither the Black queen, Black queen's bishop, nor Black queen's rook could have ever gotten out (we recall that the Black king has not moved). Therefore the pawn on f5 could not have just moved from e6 or g6, so it has just moved from f6 or f7.

Now, the piece captured by the White pawn on h3 must be the Black king's rook (the only other missing Black piece that could have been captured by a pawn is the Black king's bishop, traveling only on black squares). First the capture on h6 must have occurred to let it out (since before the last move the pawn was always on f6 or f7, and the Black king hasn't moved). The knight on g8 couldn't have gotten there via f6 (where it would have checked the Black king and made it move), so it got there via h6. This means that the knight got to g8 *before* the capture was made on h6 and has not moved any time after the capture. Hence just before the capture, the knight was on g8. Hence at that time the rook could not have been on h8 or it could never have subsequently left there, so it was on f8 or f7. Now, if the pawn on f5 had just moved from f7, then at the time in question the rook stood on f8 and could not have left there. So the pawn on f5 has just moved from f6 and not f7. At some time or other, in order to escape, the Black rook had to move from f7 to g7. At that time, the pawn on f5 was on f6, so the bishop on g7 must have been on f8 or g8. So the bishop on g7 has indeed been to the eighth row.

32 • MYSTERY OF THE LOST PAWN

The Black piece captured on c3 was not a knight, since it would have attacked the queen, nor a bishop traveling on white squares, nor a pawn (which would involve too many captures of White pieces), hence was a queen, rook, or bishop on black squares. But there are now on the board a Black queen, two Black rooks, and one Black bishop on black squares. So either it is the case that one of these is promoted, or else a promoted Black piece was captured on c3. Thus a Black pawn has promoted. It is the missing pawn from g7, and must have made one capture on either the f-file or h-file, then another on g2, and promoted on g1 (it could not have promoted via e2 because the White royalty have never been under attack). White is missing a queen's rook and king's bishop. The latter was captured on g2, since the queen's rook couldn't get there. Now, before the queen's rook got out to be captured, *both* pawns on b3 and c3 had made their captures (because before the capture on c3, the White queen's bishop was confined to c1), so both these captures were made before the promotion. So the promoted piece was not captured on c3, hence is now on the board, and is a queen, rook, or bishop. A promoted bishop couldn't have left g1, and a promoted rook on g1 couldn't get out to a8 or h8, so it is the queen who is promoted.

33 • THE LADY AND THE KNIGHT

The White queen's bishop and rook were captured on b6 and f6. The pawns on a3 and b3 have cross-captured to let the rook out (because the White king has never moved). They captured the two Black bishops. The only piece which was free to get out before any capture was the Black queen's bishop, hence it was the first to fall. It fell on b3, thus freeing the White queen's rook, but not the White queen's bishop. Now, the Black king's bishop was captured on a3, but before that the capture on f6 occurred to let it out. Also since the White queen's bishop couldn't get out till after the capture on a3, the capture on f6 occurred before the White queen's bishop got out to be captured. Hence the White queen's bishop was captured on b6 (and the rook was captured on f6). This means that the capture on a3 occurred before the capture on b6. Also the Black rook on a2 was there (or on b1, a1, or c1) *before* the capture on a3 (because after that capture, the rook could never have gotten in). So the rook on a2 got behind the pawns on a3 and b3 before the capture on b6 and hence before the Black queen's rook got out. Therefore the rook on a2 must be the king's rook.

(The sequence of moves was this: First the Black queen's bishop got captured on b3. Then the White queen's rook got out and was captured on f6. This freed both the Black king's bishop and the Black king's rook. The Black king's rook then got over to a2, then the Black king's bishop got captured on a3. Then the White queen's bishop got out and was captured on b6. Then the Black queen's rook got out and went to g2.)

34 • STORY OF THE MAGIC CARPET

Black is missing a queen, queen's bishop, and queen's rook. None of these could get out until after the capture on e6. Therefore no Black piece was captured by a pawn until after the capture on e6. In particular, the capture on e6 occurred before the capture on c3. Hence the White queen's rook was not the piece captured on e6 (because it couldn't get out yet). Also the missing White pawn was not captured on e6 because it would have had to capture at least one Black piece to get to e6 and none were out on the board to be captured. So the missing White pawn has promoted. It must have captured some Black pieces en route to promotion, hence the capture on e6 occurred before the promotion and was of an original piece. Now, if the promoted pawn came from g2, it must have captured two pieces (since it didn't cross f7 and check the king). If it came from f2 it must have made at least one capture and the pawn on f3 another. In either case the pawns originally from f2 and g2 have collectively made two captures and the pawn on c3 a third. Therefore all three missing Black pieces have been captured by pawns. In particular, the Black queen has been captured by a pawn, was under attack immediately before that, hence was never previously under attack by anything else. This means that the original White piece captured on e6 was not a knight, for right before the capture the queen's bishop hadn't yet moved, hence the queen was still on her own square and would have been under attack from the knight. It was also not a queen, since neither of the White royalty has been under attack. Nor was it the queen's rook (which we have already established), nor the king's rook, since White can castle. Therefore it was a bishop. Hence the bishop on c4 must be promoted. It must have promoted on g8, and at the time of promotion both pawns on e6 and g6 were already there. (We already know this about the pawn on e6, and the White pawn en route to promotion must have crossed the square g7, since it didn't cross f7.) Now the promoted bishop, promoted on g8, could never have *legally* left g8 to get to c4, so it is this bishop that has used the magic carpet! Before it got the magic carpet it moved legally, so could not have moved to f7. So it acquired the magic carpet on g8. Thus the bishop on c4, which is really a promoted pawn, has acquired the magic carpet on g8.

35 • THE PHANTOM BISHOP

The capture by the pawn on c6 was not of the White king's bishop (which never left f1) nor of the White queen's bishop (which traveled only on black squares), hence was of the White queen. Now, the Black bishop on a2 was there (or on b1) *before* the pawn on b3 had moved from b2. And before this, the capture on c6 occurred to let the Black queen's bishop out. So the White queen was captured on c6 while the pawn on b3 was still on b2. Now the queen couldn't get out to be captured until after the queen's bishop had either moved away from c1 or was captured on c1. It couldn't have moved away from c1 because the pawn was still on b2! Therefore it got captured on c1, thus freeing the queen. So the phantom bishop is off the board and was captured on its own square.

36 • PHANTOM BISHOP II

If the bishop on a2 is *original*, then by the same argument as the last problem, the White queen's bishop would have to have been captured on its own square. But we are now given that it wasn't captured on its own square. Therefore the bishop on a2 must be promoted. The promoting pawn must be from a7, and must have promoted on b1 (a white square) and made one capture. It must have captured the phantom bishop (the queen was captured on c6 as before). And it must have captured the phantom bishop behind the pawn on b3, hence made the capture on b2.

37 • TWO PHANTOM BISHOPS

Again the White queen was captured on c6, and since there are eight Black pawns on the board, the bishop on a2 must be original and, by virtually the same argument as in Problem 35, the White queen's bishop must have been captured on its home square. Therefore the White phantom bishop now on the board must be a promoted bishop. The promoted pawn must have come from g2, and since at most one Black piece is missing, the pawn promoted on h8. And it must still be on h8 because of the pawn on g7.

The only Black piece which the promoting pawn could have captured is the Black phantom bishop. And that capture must have occurred on h6.

38 • BEST OF THE PHANTOM BISHOPS

Black is missing the king's bishop. The queen's bishop—the phantom—may have been captured or it may be on the board though invisible. Black is missing no other pieces. White is missing the queen, queen's bishop, and queen's rook from his original set of fifteen pieces (a knight was given as odds).

Clearly the Black king's bishop was captured on a3. It couldn't have gotten out till after the capture on f6, so the capture on f6 occurred while the pawn on a3 was still on b2. The only White piece that could have gotten out to be captured on f6 while the White pawn was still on b2 was the queen, so the queen was the first White or Black piece to be captured, and she was captured on f6. Since the pawn on a3 was on b2 at all times before the queen got captured, and the White king has not moved, then the pawn on c3 must have moved from c2 to let the queen out. Therefore the pawn on c3 was there before the pawn from b2 made its capture on a3. Now, the White piece captured on a6 (a White square) was not the White queen's bishop (which couldn't have gotten out anyhow after the pawn was on c3), nor the White queen (which was previously captured on f6), hence was the White queen's rook. But the rook couldn't get out till *after* the capture on a3. Now, the knight on b1 couldn't have gotten there after the capture on a3, because the pawn on c3 was on the only other square it could have come from. Therefore, right before the capture on a3, the rook stood on c1 or c2. But the rook could not have gotten to c1 or c2 unless the White queen's bishop had first been captured on its own square! Therefore the White queen's bishop was captured on its own square, so no other piece was. In particular, the phantom bishop was not captured on its own square, hence if it got captured at all, it didn't get captured till after the capture on a6 of the White rook, which in turn was after the capture on a3, which in turn was after the knight was secure on b1. So the phantom bishop was not captured by the knight on b1, since this knight was locked in there *before* the bishop got free. Also the phantom bishop was clearly not captured by any other White piece now on the board. Nor was it captured by any missing White piece, because all missing White pieces were captured before the bishop got free. So the phantom bishop is now somewhere on the board, and is having great fun frightening all the other pieces.

39 • STORY OF THE GENIE

White has fifteen real pieces on the board and is missing the queen. Black is missing a queen, the pawn from g7, and the Black queen's rook. The Black queen's rook was captured on a7, a8, or b8. The White queen was captured on b6. Before this the pawn on c3 made its capture to let the queen out (the king hasn't moved). What did the pawn capture on c3? Not the Black queen, because it couldn't get out before the capture of the White queen on b6, and also not the Black queen's rook, nor the pawn from g7, since it had no pieces to capture. Hence this pawn promoted, coming straight down the g-file and promoting on g1. Before this, the pawn on h3 must have made its capture. Since Black is missing only two pieces eligible to be captured by pawns (the Black queen's rook was not eligible) and the queen is one of them and it was not captured on c3, the queen was captured on h3. Therefore the capture on c3 occurred before the capture on h3 (to let out the *White* queen, which got captured on b6 to let out the *Black* queen), and hence before the pawn from g7 promoted. Therefore an original piece was captured on c3. Hence the promoted Black piece is now on the board—for if it had been captured, this capture, together with the captures on c3 and h3 and the capture of the Black queen's rook, would make one too many captures. Which Black piece on the board is promoted? Not the rook on h8, since Black can castle. Also it cannot be a knight, since a promoted knight from g1 could not have escaped via h3, since the pawn was already there, nor via f3 without checking the king and making it move. So it is the bishop on g7. Thus a bishop was promoted on g1. In order for it to escape from g1, the pawn on f3 must have first moved from g2. Therefore the pawns on

153

e2, f3, and h3 have all been there ever since the bishop escaped, hence the knight on g1, if it were *real*, could never have gotten in to g1. So the knight on g1 is the genie.

40 • SECOND STORY OF THE GENIE

It is quite easy to find the Genie-Dissolving Powder. The pawn on h4 captured the Black queen's castle (on h4 or h3). So the pawns on b6 and c6 have cross-captured to let it out. The pieces captured on b6 and c6 were a knight and the king's castle. The king's castle couldn't get out till *after* the Black castle was captured by the pawn on h4, hence the first of the two captures on b6 and c6 was of the knight. Also, for the Black rook to have gotten out, the capture on c6 must have occurred before the capture on b6, because while a pawn was still on b7 the Black queen's bishop was still on c8. Therefore the knight was captured on c6 and the White castle was captured on b6.

41 • STORY OF THE INCONSPICUOUS GENIE

The problem is that the White king's rook was taken at c6 to let out the Black queen's rook, but the Black queen's rook would have to have been taken at f3 to let out the White king's rook. This can't be remedied by taking off either remaining rook, because neither of them could have gotten out on the board to be captured. Also, it won't remedy the situation to take off a knight, because if a knight was taken at either of the crucial points (f3 and c6), one of the royalty (which couldn't have moved up to that time) would have been in check. Also the White queen couldn't have gotten onto the board, and the Black queen and her bishop were stuck behind the pawn at d7. So none of these can be the genie.

The only way to relieve the difficulty is to remove the bishop on f1, so the White king's rook could have gotten out on the f-file after the pawn on e3 captured the Black king's bishop. So the bishop on f1 is the genie.
Remark: It was still the White king's rook that was captured on c6, because the White king's bishop couldn't get out till after the Black queen's rook got out and was captured on f3.

For the second problem, the analysis is as follows. We have to relieve the same problem as faced in the preceding study. If the rook on h1 is the genie, then the bishop on f1 must also be a genie for the same reason as before—i.e., if it were real, the missing rook from h1 could never have gotten out prior to the capture on f3. So the rook on h1 is real. Also it would not help if the bishop on f1 were the genie, for then the rook on h1 would be real, hence couldn't have been captured on c6, nor could the White king's bishop have left f1 to get captured on c6 prior to the capture on f3. So neither of the pieces on f1 and h1 is a genie. Also none of the pieces on a1, b1, and d1 is the genie for the same reasons as in the preceding problem. Thus all the White pieces shown on the board are real, and the White piece captured on c6 must have been a knight. Now, the Black queen shown on d8 can't be a genie, for then the real Black queen would have been captured and thus would have been under attack before she was captured. So the Black queen on d8 is real. How would she have had a square to stand on to avoid attack while the White knight was on c6 and a Black pawn on d7? This is possible only if the Black bishop shown on c8 is the genie; the queen could then have stood on c8. So the Black bishop on c8 is the genie.

Unless Black can castle, White has a mate in two by moving his queen to a6; then regardless of Black's next move, queen to c8 mates Black. Thus Black can save himself only by castling. Can he castle? We prove that if the bishop on h2 is real, then he cannot. Equivalently, we prove that if Black can castle, then the bishop on h2 is a genie.

So suppose Black can castle. Then the Black queen's rook never got out to be captured by a pawn. In particular, the piece captured on c3 was not the Black queen's rook. It also was not the missing Black pawn, because if the pawn were from e7 it would have made two captures and the pawns on f6 and g6 two more, which is two too many, and if it were from any other square even more captures would be involved. Therefore the missing Black pawn promoted.

The White pawn on e4 comes from e2, otherwise it and the pawns on c3 and g3 would have made two more captures than is possible, and this pawn has always been on the e-file, for otherwise it would have made two captures and the pawn on c3 a third, which is one too many, considering that the Black queen's rook could not have gotten out to be captured by either of them. Therefore if the promoted Black pawn came from e7 it would have to have made at least one capture and the pawns on f6 and g6 two more, which is one too many. Therefore the Black pawn that promoted did not come from e7. So the pawn on f6 comes from e7 and the promoted pawn comes from g7 or h7. Now, the White pawn on g3 cannot come from h2, or it plus the pawn on c3 would have made two captures, both on black squares, which is not possible, since the Black queen's bishop (traveling on white squares) would have to be one of the pieces captured. So the pawn on g3 comes from g2. This means that the Black promoted pawn couldn't have come from g7, since it would then have had to have made a capture, the pawn on g6 another (from h7), and the pawn on f6 a third. So the Black promoting pawn comes from h7.

Now, the White pawn from h2 has also promoted, for the White piece captured on f6 was not the White king's bishop, nor could it have been the pawn from h2, since it would have had to have made two captures to get there and, with the capture by the pawn on c3, this is too many. Thus both pawns from h2 and h7 have promoted. One of them must have captured first to let the other by. It couldn't have been the White pawn, for that could have happened only on g7 while the Black pawn was on h7, but g7 like c3 is a black square. So it was the Black pawn which made a capture from h3 to g2 while the White pawn was still on h2. Thus the pawn on g3 was there *before* the pawn from h2 moved. This means that the bishop shown on h2 must be a genie, because if it were real, it could not have gotten to g1 or h2 either before the pawn from h2 moved nor after the pawn on g3 was there.

We have thus proved that if Black can castle, the bishop shown on h2 is a genie. Stated otherwise, if the bishop on h4 is the genie, then the bishop on h2 is not and Black cannot castle. So if the bishop on h4 is the genie, then White can win the game in two moves.

43 · STORY OF THE TRANSFORMED BISHOP

The White pawns on b3 and f4 have collectively made three captures. Therefore there are at most thirteen Black pieces on the board. Since fourteen are shown, then it must be that one of the Black knights is really a White bishop. Now to find out which one.

White is missing the White queen's rook and one bishop, which were captured by the pawns on e6 and h6. Therefore neither White bishop could have been captured on its own square. Since the White queen's bishop was not captured on its own square, and the White queen has not moved, then the White queen's castle did not get out till after both the capture on b3 and the capture on e3 by the pawn from d2 (to first let out the queen's bishop). Now, the three Black pieces captured by the White pawns on b3 and f4 consist of a knight, the king's bishop, and the king's rook. Neither the bishop nor the rook could get out before the capture on h6. Therefore only one Black piece (the knight) could have been captured by a White pawn before the capture on h6. Since the White queen's rook couldn't get out till after the White pawns made two captures, it was not the piece captured on h6. So the White queen's bishop was captured on h6, and the White queen's rook subsequently on e6. (The sequence was this: First a Black knight was captured on e3, then the White queen's bishop got out and was captured on h6, then the Black king's bishop and the Black king's rook could get out. The rook was then captured on b3, then the White queen's rook got out and was captured on e6. The Black king's bishop was captured some time or other at f4—sometime after the capture on h6.)

Since the White queen's bishop was captured on h6, then it must be the Black knight shown on c6 which is really the White bishop.

44 • STORY OF THE ENCHANTED HORSE

The Black promoting pawn came from d7. It must have made two captures and promoted on b1. The pawn on b3 comes from b2 and must have moved from b2 prior to the capture on b2, hence prior to the promotion. Thus the knight on a1 is certainly original (since it was on a1 prior to the pawn on b3 being there). Also the promoting Black pawn made two captures, one of which was of the White queen's bishop. The other couldn't have been of the pawn from a2, hence this White pawn also promoted. Since no promoted White pieces are now on the board, then the promoted White piece was one of the two pieces captured by the Black promoting pawn. So the White pawn promoted before the Black pawn. The White pawn started from a2, went to a6 and captured the Black queen on b7, and then promoted on b8. So the pawn on b6 was there *before* the White pawn promoted, and hence also before the third Black knight was made. So the knight on a8 must also be original (since again a knight couldn't get to a8 after the pawn on b6 moved). Therefore the knight on h8 is the promoted knight.

For the twin problem, we first observe that since the White king never moved, the Black pawn couldn't promote on e1. So again it promoted on b1 or d1 after making two captures. Then for the same reason as before, the knight on a1 must be original. This time, the pawn from e2 must have promoted to a piece captured by the Black promoting pawn. The White promoting pawn must have made a capture on the d-file of the Black king's bishop. Before this, the pawn on g6 moved to let it out. So the knight on h8 must be original. And this time the knight on a8 is the promoted knight.

Neither of the White pieces captured by the pawns on c6 and d6 were knights, since the Black royalty has never moved or been under attack. Also the White king's bishop was captured on its own square. So one of the two pieces captured on c6 and d6 was a rook, and either the other was the missing pawn from a2 or else this pawn promoted. The former is not possible, for the following reasons.

The pawn from a2 certainly couldn't have captured three pieces to get to d6, since Black is missing only three pieces and one of them was captured on g3. Now suppose it was captured on c6. Then it captured two pieces *before* the capture on c6, hence it captured two pieces before the Black queen's bishop could get out, which would leave the Black queen's bishop as the only piece left which could be captured on g3, which is not possible, since g3 is a black square. Therefore the pawn from a2 promoted, and is either now on the board or else was one of the two pieces captured on c6 or d6.

The promoting pawn from a2 obviously promoted on a8 after making two captures (one on the b-file and the second on a7). It didn't promote to a rook because a rook couldn't have gotten out onto the board (to be at h1 or to have been captured at c6 or d6). Likewise a bishop couldn't have left a8. So the pawn promoted to a queen or knight. But it couldn't be a knight because the knight would have to be either now on the board, which it isn't, or else one of the pieces captured on c6 or d6, which is not possible since neither of the Black royalty has moved or been under attack. Therefore the pawn definitely promoted to a queen.

The crucial question now remains: Was it the promoted queen or the original queen which was captured on c6 or d6? Well now, the promoting pawn from a2 captured two pieces en route to promotion, and one of them must have been the Black queen's bishop, since this bishop was not the piece captured on g3. Therefore the capture on c6 occurred prior to the promotion, hence the promoted queen was not captured on c6. Could it have been captured on d6? (This is the most interesting part of the proof!) Suppose it were; consider the position at the time the White queen stood on d6 just about to be captured by the pawn then on e7. The pawn on c6 had already made its capture, so unless some piece stood on d7, the Black queen (which has never moved) would have been attacked by the White queen. But we are given that the Black queen was never attacked. So some piece stood on d7. But what piece could it have been? The pawn on d6 then stood on e7, so the Black king's bishop hadn't moved out yet, so the two Black pieces previously captured by the promoting White pawn were the Black queen's bishop and the missing knight. So neither of these two pieces then stood on d7. Also the Black king's knight has made only one move, hence it never stood on d7. Also no Black rook has ever been on d7. Therefore no Black piece stood on d7 at the time. Nor could it have been a White piece, since it couldn't have been a rook or the original queen, which would have attacked the Black queen, nor the White king's bishop, which never left f1, nor a knight, because no White knight has gone beyond the sixth row.

This proves that it is impossible that the promoted White queen was captured, hence she is now on the board standing innocently and coyly next to Haroun. So Haroun's suspicions were fully justified.

46 · WHICH QUEEN?

Black is missing both rooks and the king's bishop; White is missing both bishops and the king's rook. The first of these six pieces free to get out prior to any capture was the Black king's bishop. It then got captured on e3, thus freeing the White queen's bishop. Now the White queen's bishop couldn't have been captured on g6, so it was the White king's bishop or White king's rook which was captured there. But this did not happen till after the capture on h3 (to let the rook or bishop out). What was captured on h3? Not the Black king's rook, because it couldn't have gotten out yet ("yet" meaning before the capture of the White king's bishop or White king's rook on g6). Hence it was the Black *queen's* rook that was captured on h3, so the pawns on a6 and b6 have cross-captured to let it out. The sequence was this: After the pawn from d2 captured the Black king's bishop on e3 the White queen's bishop got out and was captured on b6. Then the Black queen's rook got out and was captured on h3. Then the White king's bishop and the White king's rook got out and one of them was captured on a6 and the other on g6. As to the Black *king's* rook, it must have been captured on b7 by the pawn from c2 en route to promotion on b8 to a queen. Clearly the pawn on a6 made its capture before the White pawn captured on b7, hence before the promotion. Also the pawn on b6 made its capture before the promotion, since it had to let out the Black queen's rook to get captured on h3 to let out a piece to be captured on g6 to let out the Black king's rook which was captured on a6. Also the pawn on e6 moved from e7 to let out the Black king's bishop before any captures were made at all. Also the pawn on d5 was there from Black's first move. Therefore all five pawns on a6, b6, c7, d5, and e6 were on their present squares before the promotion. This means that there was no way the promoted queen could have gone from b8 to g8 without checking the Black king, because it could only get out by standing on c6, and no piece then stood on d7 to block check, since we are given that d7 has never been reoccupied. Therefore the queen on b8 must be the promoted one and Amelia stands on g8.

<h2>47 · A NEW COMPLICATION</h2>

Black is missing both rooks and both bishops, and White is missing one rook and both bishops. Since the pawn on d4 comes from d2, the promoting White pawn must come from e2. It promoted either on b8 or h8 and made three captures in either case; the fourth missing Black piece was captured on h3.

We prove that, as in the preceding problem, the pawns on a6 and b6 have cross-captured, though this time for a rather different reason: Suppose they hadn't cross-captured. Then the Black queen's rook which was one of the pieces captured by the promoting pawn could never have gotten out, hence would have to have been captured on b7. But this is impossible, since the pawns on d4 and d5 were there from the first moves of the game and the only way the promoting pawn could get from e2 to the c-file is by making its captures on d3 and c4, which would mean that all four missing Black pieces were captured on white squares. This can't be, since the Black king's bishop is among the missing pieces. Therefore the pawns on a6 and b6 did cross-capture, and the three missing White pieces were captured on a6, b6, and g6.

The piece captured on h3 was the queen's bishop or one of the castles. Of these three pieces together with the three missing White pieces, the first one to be free to move prior to any capture is the White queen's bishop (because the bishops on white squares have never traversed e2 or d7; they got out in the "fianchetto" manner— i.e., via the side squares g2 and b7). The White queen's bishop was captured on b6. Then (as in the last problem) the Black queen's rook got out and was captured on h3, thus freeing the White king's bishop and White king's rook to be captured on the squares a6 and g6. So again, the pawn on b6 made its capture before the pawn on a6. This means that the promoting White pawn couldn't have promoted on b8, because it would have had to make a capture on the white square b7 (as well as d3 and c4), so all four missing Black pieces would have been captured on white squares. Hence the promotion occurred on h8.

As in the last problem, all pawns on a6, b6, c7, d5, e6, and on g6 as well were on their present squares prior to the promotion; the ones on a6 and e6 to let out the bishops (to be captured by the promoting pawn), the one on g6 to let out the Black king's rook, the one on d5 was there after the first move, and the one on b6 was there before the one on a6, as we have already proved. So these pawns were all there before the promotion. Now, in the last problem, the promoted queen could not get *out* from b8 to g8 without checking the king; in this problem the promoted queen could not get from g8 *into* b8 without checking the king, since again no piece has stood on d7 since the pawn left that square. So this time the queen on g8 is the promoted one and Amelia stands on b8.

48 • THE RESCUE OF AMELIA

Since the White queen was captured by a rook, the pieces captured by the pawns on c6 and h6 are the two White bishops. The first piece free to be captured by any of the four capturing pawns (c3, c6, f3, h6) was the Black queen's bishop, and it got captured on f3. Then the White king's bishop got out and must have been captured on c6. This let out the Black queen's rook, which then got captured on c3. (This let out the White queen's bishop, which then got captured on h6.) Now before the Black queen's rook got captured on c3 the White queen couldn't have been captured by a rook, because the queen couldn't get out nor could a rook get in. Therefore the White queen got captured *after* the Black queen's rook did, hence was captured by the king's rook. So Amelia is imprisoned in the castle on the east border.

49 • ADVENTURE IN THE FOREST

The pawns on d6 and e6 have obviously cross-captured to let the White rook in to d7. Since the rook never stood on e7 (where it would have checked the Black king and made him move), it must have come to d7 via d6, hence the pawn on e6 made its capture before the pawn on d6.

The promoting White pawn comes from e2 and must have promoted on h8 or b8 after capturing all three missing Black pieces. One of these pieces was the Black king's bishop, which couldn't get out till after the capture on d6. And we know that the capture on e6 occurred before that, hence *both* captures on d6 and e6 occurred before the promotion. Hence the rook on d7 is one of the original ones. Also, the promotion was on h8 rather than b8, because a promoted rook on b8 could never have gotten out to a1 or h1.

Now, the White pieces captured on d6 and e6 before the promotion are the two bishops. Before the queen's bishop was captured on

161

d6 the pawn on b3 had moved to let it out. After the pawn on b3 was there, a promoted rook could never have gotten in to a1. So the promoted White castle stands on h1 (and got there while the pawn on g3 was still on g2 and after the pawn on h4 was there).

50 · IN SEARCH OF THE WATER OF LIFE

Black is missing only one pawn, and this pawn could never have been captured by the pawn on b4, since it would have to have made too many captures to get there. Hence it promoted. It must be the pawn from g7, since if it were from e7, it would have to have made two captures to promote, and the pawn on e5 two more to get from g7. So the pawn from g7 promoted, first making one capture to get to the f-file, then another on e2, promoting on e1 after the White king moved to d2. The promoted piece either got captured by the pawn on b4 or else replaced the piece captured by that pawn. The Black pawn couldn't have promoted on e1 to a queen or bishop, since it would have checked the king on d2 and made it move a second time. Nor could it have been a knight, which could have escaped only by f3, which would also have checked the king on d2. So it promoted to a rook. This rook could have escaped only by going to a1 and then out on the a-file. Therefore the White rook on a1 has moved. Also the Black rook could escape only *after* the capture on b3 or b4 by the White pawn, hence this pawn captured an original rook and the promoted rook is on the board. It could not have gotten into a8, since the Black royalty have not moved, so it now stands on h8. Thus the rooks on a1 and h8 have both moved. We are given that two of the rooks didn't move, so they must be the ones on h1 and a8.

• PROBLEMS OF KAZIR'S COURT •

K1

1.	S-b3	K × S
2.	S-c5 Mate	

K2

1. P-c8 = B
   If K-c4 then B-e6 Mate
   If K-c6 then B-b7 Mate

K3

1.	P-b4	P × P e.p.
2.	B-c7 Mate	

K4

1.	B-a8	P-h4
2.	P-b7	K-h1
3.	P-b8 = Q or B Mate.	

K5

1. R-h7

   (a)   R × R on h7
2. R × R Mate.

   (b)   R-g6 check
2. R (g1) × R   R × R or R-g8
3. R-g8 or R × R Mate.

   (c)   R (g8) moves
2. R (h7) × R check   R-g8
3. R × R Mate.

K6

1.	B-h2 (threat: P-g4, mate)	P × P
2.	P-g4 check	K-e6
3.	P-f5 check	K-d7
4.	P-e6 check	
	(a)	K-c6
5.	b-e8 mate	
	(b)	K-c8
6.	P-d7 Mate.	

K7

(1)

1.	Q × R check	S × Q check
2.	R-g7 check	R-e7 Mate

(2)

1.	R (e7) × R double check	K-f7
2.	R-e7 check	R × R Mate.

K8

1.	S-f4 check	K-g3
2.	S-d3 check	K-h3
3.	S-f2 check	S × S Mate.

K9

1.	Q-h7 check	S-h6
2.	Q × S check	B-h5
3.	Q-g5 check	S × Q
4.	B-e1 check	P-g3 Mate.

K10

1.	P-b6 check	K-a6
2.	B-b5 check	K × S
3.	P-b4 check	P × P e.p.
4.	B-b2	B-b1 Mate.

K11

1.	R × P check	R-f6
2.	R-g7 check	R-d6
3.	R-b7 check	R-f6
4.	R-b2	P (either) × R Mate.

K12

1.	B-a6 check	
	(a)	P × R = Q
2.	S-g3 check	Q × S Mate.
	(b)	P × R = R
2.	Q-e3 check	R × Q
3.	P-d3 check	R × P Mate.
	(c)	P × R = S
2.	B-d3 check	S × B
3.	S-c5 check	S × S
4.	Q-e6 check	S × Q Mate.
	(d)	P × R = B
2.	S-g3 check	B × S
3.	Q-e6 check	B-e5
4.	Q-f5 check	P × Q
5.	S-f6 check	P or B × S Mate.

AFTERWORD

I have received many interesting letters concerning my Holmes book. Some correspondents asked me whether the problems in the book were original. The answer is yes, all the problems in that book, as well as all the problems in this book, are original. Other correspondents expressed curiosity about how I ever got into retrograde analysis in the first place. Well, the story behind that, as well as how my chess books got written, is an interesting one.

I composed my first chess problem in 1925, when I was sixteen; it was a conventional two-mover. I showed it to several of my older friends. One of them said: "Hmm! Now, if *I* were to compose a chess problem, it would be of a very different sort!" "Of *what* sort?" I asked. "Well," he replied, "if I were to compose a chess problem, it would not be in the usual form— White to play and mate in so many moves—rather, it would be to deduce what had happened earlier in the game." This struck me as a fascinating idea, and I straightaway set to work and composed a problem in retrograde analysis (the "Mystery of the Missing Piece" in the Holmes volume). I had not yet heard the phrase "retrograde analysis," nor did I even suspect that there was such a field. Retrograde analysis was hardly known in the United States, having been developed by a small coterie of problem composers in England and Europe. Anyhow, my friends were extremely enthusiastic about the problem, and everyone encouraged me to compose more, which I have been doing ever since at various odd stages of my life. Because there were no American newspapers or periodicals that published retrograde problems, and because I knew of no foreign periodicals that did, most of my problems remained unpublished.

Sometime in the forties, it occurred to me that this type of problem was ideal for incorporation into stories! And, inspired by Lewis Carroll, I conceived the notion of stories in which the chess pieces themselves should be the *dramatis personae*. The Arabian Knights somehow leaped to mind as the proper setting—Haroun Al Rashid would be the White king, his grand vizier would be the White king's bishop, and so on. At that time I wrote a few of these "Arabian Knight's" Tales, and planned to make a whole book of them one day. When this nebulous "one day" would be, I had no idea, and it might never have been, except for the following remarkable set of circumstances.

Early in 1957, I showed the opening problem of this book, "Where Is the White King?" (as well as many of my other problems), to several fellow graduate students at Princeton. A distinguished logician then visiting the Institute for Advanced Studies was also present. One of the students said to me: "Smullyan, why don't you publish this problem before somebody else does?" I laughed and naively replied, "Why would anybody want to do a thing like that?" I forgot about the matter until a few weeks later, when I met the logician from the Institute, who said: "Hey, Smullyan, how come your problem was just published in the *Manchester Guardian* without crediting you as the author?" I went immediately to the graduate student and asked him if he knew anything about this. "Oh yes," he replied, "I showed the problem to my father, who has had frequent correspondence with the *Guardian*'s chess editor, and he sent him your problem with the comment: 'Instead of the usual type of chess problem, why don't you publish this one?'" I was, of course, gratified that my problem had been published, but I naturally expressed disappointment that my authorship had not been mentioned. "Oh, I'll speak to my father about that," he said. A couple of weeks later, I received a very nice letter from the chess editor of the *Manchester Guardian*, expressing regret that he had not known I was the author of this "delightful work," and assuring me that my authorship would be acknowledged in the next issue. He also asked if I had other problems of this genre that I could send him, and in the next year I published several problems in the *Manchester Guardian*, which in turn led to the publication of a few others in European journals as well as in the Canadian *Chess-Chat*.

No further relevant incidents occurred until the early seventies. Then, in March 1973, the same problem that had first been published in the *Manchester Guardian* came out in Martin Gardner's column in *Scientific American!* It had been sent in by a correspondent, with a note saying that, while he found it remarkable, he did not know who had originated it. As luck would have it, I missed that issue of *Scientific American*, but a friend saw it and promptly wrote to Martin Gardner that the problem had been devised about twenty years before by Raymond Smullyan, and that it was one of a large collection of unpublished chess problems invented by Smullyan when they were both students at the University of Chicago.

This incident led to a happy renewal of my acquaintance with Martin Gardner, who urged me to stop dilly-dallying around and get the book written! Well, I have to admit, a couple of years went by. Then, on sabbatical leave for a semester, and prodded by a publisher's interest (not the one which eventually brought out the books), I decided to spend my entire time on the project, which I did. At first, I planned to put all my puzzles into the "Arabian Knights" book, but fate intervened. The well-known retrograde expert and composer Mannis Charosch (one of the few in America at that time) had seen some of my problems, and kindly sent me a copy of his excellent paper "Detective at the Chessboard" (*Journal of Recreational Mathematics*, Vol. 5, Nov. 2), which is a fine introduction to retrograde analysis for the general reader. The title "Detective at the Chessboard" instantly captured my fancy, and I thought: "Why not have a *real* detective at the chessboard, and while I'm at it, why not Sherlock Holmes!" So I changed my plan and decided to divide the problems into *two* books: one on Sherlock Holmes and the other on the Arabian

Knights. (I took great care to divide the best of my problems evenly between the two books.) And that's how the books came to be written.

I wish to thank my correspondents—far too numerous to list individually—for their extremely helpful suggestions for improving further editions of my Holmes book. My belated thanks to Dr. Jack Kotik for his intriguing motto, "To Know the Past One Must First Know the Future," which I used in one of my Holmes stories.

For the present volume, I am grateful to Bill Snead of Amarillo, Texas, Professor Andy Liu of the University of Alberta, and Professor Robert Kurtz for help with correcting earlier drafts. To Greer Fitting goes my enthusiastic appreciation for the very charming illustrations that adorn the book. Once again, it has been a real pleasure to work with the editorial staff of Alfred A. Knopf; special thanks are due to Melvin Rosenthal for skillfully guiding the book through the successive stages of proof, and to Virginia Tan for her adept handling of the complex problems of design and layout. And, above all, I wish to thank my editor, Ann Close, for the great care she bestowed on the manuscript.

Raymond Smullyan

Elka Park, New York
January 1, 1981

OXFORD

MORE OXFORD PAPERBACKS

Details of a selection of other Oxford Paperbacks follow. A complete list of Oxford Paperbacks, including The World's Classics, Twentieth-Century Classics, OPUS, Past Masters, Oxford Authors, Oxford Shakespeare, and Oxford Paperback Reference, is available in the UK from the General Publicity Department, Oxford University Press (RS), Walton Street, Oxford, OX2 6DP.

In the USA, complete lists are available from the Paperbacks Marketing Manager, Oxford University Press, 200 Madison Avenue, New York, NY 10016.

Oxford Paperbacks are available from all good bookshops. In case of difficulty, customers in the UK can order direct from Oxford University Press Bookshop, 116 High Street, Oxford, Freepost, OX1 4BR, enclosing full payment. Please add 10 per cent of the published price for postage and packing.

SCIENCE IN OXFORD PAPERBACKS

Oxford Paperbacks' expanding science and mathematics list offers a range of books across the scientific spectrum by men and women at the forefront of their fields, including Richard Dawkins, Martin Gardner, James Lovelock, Raymond Smullyan, and Nobel Prize winners Peter Medawar and Gerald Edelman.

THE SELFISH GENE

Second Edition

Richard Dawkins

Our genes made us. We animals exist for their preservation and are nothing more than their throwaway survival machines. The world of the selfish gene is one of savage competition, ruthless exploitation, and deceit. But what of the acts of apparent altruism found in nature—the bees who commit suicide when they sting to protect the hive, or the birds who risk their lives to warn the flock of an approaching hawk? Do they contravene the fundamental law of gene selfishness? By no means: Dawkins shows that the selfish gene is also the subtle gene. And he holds out the hope that our species—alone on earth—has the power to rebel against the designs of the selfish gene. This book is a call to arms. It is both manual and manifesto, and it grips like a thriller.

The Selfish Gene, Richard Dawkins's brilliant first book and still his most famous, is an international bestseller in thirteen languages. For this greatly expanded edition, endnotes have been added, giving fascinating reflections on the original text, and there are two major new chapters.

'learned, witty, and very well written . . . exhilaratingly good.' Sir Peter Medawar, *Spectator*

'Who should read this book? Everyone interested in the universe and their place in it.' Jeffrey R. Baylis, *Animal Behaviour*

'the sort of popular science writing that makes the reader feel like a genius' *New York Times*

Also in Oxford Paperbacks:

The Extended Phenotype   Richard Dawkins
The Ages of Gaia   James Lovelock
The Unheeded Cry   Bernard E. Rollin

RELIGION AND THEOLOGY
IN OXFORD PAPERBACKS

Oxford Paperbacks offers incisive studies of the philosophies and ceremonies of the world's major religions, including Christianity, Judaism, Islam, Buddhism, and Hinduism.

A HISTORY OF HERESY

David Christie-Murray

'Heresy, a cynic might say, is the opinion held by a minority of men which the majority declares unacceptable and is strong enough to punish.'

What is heresy? Who were the great heretics and what did they believe? Why might those originally condemned as heretics come to be regarded as martyrs and cherished as saints?

Heretics, those who dissent from orthodox Christian belief, have existed at all times since the Christian Church was founded and the first Christians became themselves heretics within Judaism. From earliest times too, politics, orthodoxy, and heresy have been inextricably entwined—to be a heretic was often to be a traitor and punishable by death at the stake—and heresy deserves to be placed against the background of political and social developments which shaped it.

This book is a vivid combination of narrative and comment which succeeds in both re-creating historical events and elucidating the most important—and most disputed—doctrines and philosophies.

Also in Oxford Paperbacks:

Christianity in the West 1400–1700   John Bossy
John Henry Newman: A Biography   Ian Ker
Islam: The Straight Path   John L. Esposito

MUSIC IN OXFORD PAPERBACKS

Whether your taste is classical or jazz, the Oxford Paperbacks range of music books is in tune with the interests of all music lovers.

ESSAYS ON MUSICAL ANALYSIS
Donald Tovey

Tovey's Essays are the most famous works of musical criticism in the English language. For acuteness, common sense, clarity, and wit they are probably unequalled, and they make ideal reading for anyone interested in the classical music repertory.

CHAMBER MUSIC

Chamber Music contains some of Tovey's most important essays, including those on Bach's 'Goldberg' Variations and *Art of Fugue*, and on key works by Haydn, Mozart, Beethoven, Schumann, Chopin, and Brahms.

CONCERTOS AND CHORAL WORKS

Concertos and Choral Works contains nearly all the concertos in the standard repertory, from Bach's for two violins to Walton's for viola—fifty concertos in all. The choral works include long essays on Bach's B minor Mass and Beethoven's Mass in D, amongst other famous works.

SYMPHONIES AND OTHER ORCHESTRAL WORKS

Symphonies and Other Orchestral Works contains 115 essays: on Beethoven's overtures and symphonies (including Tovey's famous study of the Ninth Symphony), all Brahms's overtures and symphonies, and many other works by composers from Bach to Vaughan Williams.

Also in Oxford Paperbacks:

Singers and the Song   Gene Lees
The Concise Oxford Dictionary of Music 3/e
Michael Kennedy
Opera Anecdotes   Ethan Mordden